# *Scientia* in Early Modern Philosophy

# STUDIES IN HISTORY AND PHILOSOPHY OF SCIENCE

VOLUME 24



For further volumes:
http://www.springer.com/series/5671

# *SCIENTIA* IN EARLY MODERN PHILOSOPHY

## SEVENTEENTH-CENTURY THINKERS ON DEMONSTRATIVE KNOWLEDGE FROM FIRST PRINCIPLES

Edited By

TOM SORELL

G.A.J. ROGERS

and

JILL KRAYE

Springer

*Editors*
Tom Sorell
University of Birmingham
Edgbaston
United Kingdom
t.sorell@bham.ac.uk

Prof. G.A.J. Rogers
University of Keele
Department of Philosophy
Keele, Staffs
United Kingdom ST5 5BG
pia04@keele.ac.uk

Prof. Jill Kraye
University of London
School of Advanced Studies
Warburg Institute
Woburn Square
London
United Kingdom WC1H 0AB

ISSN 0929-6425
ISBN 978-90-481-3076-4 e-ISBN 978-90-481-3077-1
DOI 10.1007/978-90-481-3077-1
Springer Dordrecht Heidelberg London New York

Library of Congress Control Number: 2009937293



Printed on acid-free paper

Springer is part of Springer Science+Business Media (www.springer.com)

# Contents

# Introduction

In early modern philosophy *scientia* is an honorific term. It refers to knowledge or understanding of truths in the light of principles or causes. *Scientia* is systematic knowledge of truths, truths "deducible" from principles. It is not simply knowledge-that, but knowledge-why, and not simply knowledge-why, but knowledge-why that unifies whole classes of truths known. Again, *scientia* is not merely knowledge why truths *happen* to be true. Instead, it is knowledge that the relevant truths cannot *but* be true given the relevant causes or principles. So it is knowledge of truths within a framework that makes their truth look necessitated by the underlying principles. Described like this, *scientia* is an ideal of both pre-modern and early modern philosophy. There are rival scholastic and post-scholastic interpretations of "knowledge", "cause", "principle" and "necessitated" that make this way of talking about *scientia* fit into each tradition. It took a revolution to make one interpretation supersede the other. Historiographers of the early modern period have tended in recent years to claim that it was a quieter revolution than was once thought, with many more survivals from the superseded tradition than used to be acknowledged, and there is some evidence for this in the current volume; but the reconceptualization of *scientia* is a focal point of early modern philosophy.

Before the revolution, *scientia* was knowledge of the properties of members of natural kinds based on knowledge of the essences of natural kinds. The properties were known by unaided observation. The essences were known by Aristotelian induction. They dawned on one over time with repeated observational exposures to specimens of a kind, and they were expressed by definitions. Scientia located natural kinds in an order of more to less general, which syllogistic explanations of particular explanada recapitulated. Again, *scientia* related observed facts to the unfolding of a characteristic natural history of a natural kind. The uncharacteristic or exceptional was not a possible object of *scientia*. Or, in other words, the causes relevant to *scientia* were formal and final, not efficient. The ultimate constituents of nature—the elements—were qualitatively defined. Pre-revolutionary *scientia* did not set out to get underneath or above observation, and did not set out to challenge or refine ordinary descriptions of things observed. Instead, it sought to systematize ordinary qualitative observation and description, referring it to definitions per genus et differentia that were no less observational than what they explained.

After the revolution, essences were not relative to natural kinds but to more elemental things, notably matter in its more general forms. To have *scientia* concerning celestial or terrestrial objects was to have knowledge of the motions, shapes, positions, and numbers of their parts. Explanations might invoke particles of matter that eluded observation, and might exploit both arithmetical and geometrical quantity. So post-revolutionary explanations were open both to greater informativeness and greater precision than their predecessors. At the same time they had greater generality. Because they specialized in efficient causes, the explanations were in principle empowering, indicating what interventions might, at least in principle, produce observed effects. The forms of matter being relatively few, the new explanatory principles were less complex and *ad hoc* than traditional ones. Post-revolutionary *scientia* was in all these ways thought to be an advance on what preceded it. Believers in post-revolutionary *scientia* disagreed over the nature of the faculties that gave access to the elemental explanatory principles. In Descartes it is a finite mind in the same general class as the divine mind. In Hobbes, it is a corporeal, sense-based intelligence with language-based powers of general thought. In Spinoza it is a sort of compromise between a Hobbesian and a Cartesian *res cogitans*.

*Scientia* is not what we now call "science", though what we now call "science" is closer to the conception of *scientia* of the non-Aristotelians or the anti-Aristotelians than the Aristotelians themselves. Even so, "closer" does not mean "close". As Daniel Garber points out in the opening essay, early modern practitioners of natural philosophy or physics would have considered themselves engaged in something different from "mixed mathematics" and "engineering", notwithstanding taxonomies of learning like Bacon's in the *Advancement of Learning* that insisted on affinities between them. By the same token, natural philosophy in the image of its seventeenth century practitioners is distant from the mathematized natural sciences and their related technology disciplines today. One of the current UK bodies for funding research is called "The Engineering and Physical Sciences Research Council". This way of labelling its neck of the woods in the funding world is the product of a more Baconian way of mapping than Garber finds commonplace in the early seventeenth century.

Garber's thesis is that, the more faithful we are to the self-understanding of intellectuals at the beginning of the early modern period, the less unified the thing that was called "science" then appears. Academically, it was assigned to university Arts Faculties; it was taught but not researched at universities. It was both wider and narrower in subject matter than it might be today. As the seventeenth century progressed, the self-image of practitioners of "science" changed, reflecting a sense of marked progress brought about by virtuoso talents as well as a process of unification in the sciences. Mathematics, natural philosophy and natural history all come together. Learned societies start to be formed, and the first scientific journals were founded. Almost as important, something closer to what we now call "philosophy" started to be decoupled from "science."

Stephen Gaukroger associates *scientia* mainly with the pre-revolutionary conception of explanation, and its associated problems. Mechanism is the rival, post-revolutionary explanation, and Gaukroger reviews the various anti-Aristotelian purposes it served in the hands of different moderns, starting with Bacon, who wanted to revise the Aristotelian understanding of artefact to include a human intervention in nature. Gaukroger distinguishes between post-Aristotelian matter theory and post-Aristotelian mechanistic explanation. These fused when the primary/secondary quality distinction was married with the thought that causation in nature was primarily a matter of impacts at the level of atoms. The result—foundationalist mechanism—was not the only response to Aristotle, Gaukroger claims. Although it had distinguished supporters –Descartes, Hobbes, Huygens, and Malebranche—so did a rather different approach, geared to accommodating experimental results, like observations of the effects produced by Boyle's airpump. These observations were not necessarily the effects of microscopic objects, and, according to Gaukroger, hypotheses at that level did not necessarily illuminate them. The unifying and reductive tendencies of foundationalist mechanism made sense as a response to the failings of Aristotelianism, but a looser, more *ad hoc* approach might have been justified by the range of phenomena up for explanation.

Anti-Aristotelianism did not always take the form of a self-consciously new philosophy. Sometimes it consisted of the revival and adaptation of unAristotelian but ancient sources. Epicurus was put to this use by Gassendi. Catherine Wilson explores this Epicurean vein of early modern thought, stressing its connection to questions about mortality and immortality. Her essay covers not only Gassendi, but Hobbes and Walter Charleton. One of the possibilities afforded to philosophers by the rival of ancient and (from the point of view of seventeenth century Western Europe) pagan doctrines, was detachment from theological orthodoxies, including the orthodoxy that the soul was immortal. Could not these orthodoxies be self-serving bits of ideology, put about by priests? The question could at least be posed by Charleton and (on some natural readings) Hobbes. Those who seemed to embrace mortalism, however, made room for ad hoc resurrection if God willed it. This hedging of bets notwithstanding, a Gassendist scepticism about the soul was in the air when Locke published the *Essay*. Although Locke was not himself a Gassendist, according to Wilson, his critics were also critics of Epicureanism. In short, an "earlier ideal science—appearance or idea-based in epistemology, atom-based in ontology, careless of logical relations but wedded to mechanical accounts—was the object of recovery efforts by a whole range of seventeenth century philosophers".

There was more than one strand of anti-Aristotelian philosophy of science in the early modern period, and we are sometimes blinded to this fact by the requirements of narrating a neat history of the early modern period. The neat history mentions Bacon, Descartes, Newton and other familiar figures, but omits many other intellectuals, some of whom were not in the least on the fringes of scientific life in their day, One of these, Joachim Jungius, is the subject of Stephen Clucas's essay. Clucas expounds Jungius' notion of "intellectual experience", a kind of informed and disciplined, methodologically regulated observation characteristic of astronomers, anatomists and physicians. He calls attention to the importance of this form of experience and also of forms of reasoning that started and ended in approximately true propositions. "Beginning with an essentially Aristotelian conception of *scientia*, and retaining the Aristotelian logical superstructure of his scientific induction," Clucas writes, "Jungius brought to it his experience as a student of medicine, mathematician, natural historian and chemical experimenter, and sought to extend it."

Clucas claims that Jungius's reform of Aristotelian *scientia* is in some ways comparable to Descartes's. And among the early moderns who reflected on scientia, it is Descartes who is most prominent in this volume. At the heart of the Cartesian conception is the connection between *scientia* and metaphysical certainty. To have *scientia* in Descartes's sense is to be immune from large-scale doubt, even doubt based on the intellectually destabilizing hypothesis of general human intellectual weakness or demonic deception. *Scientia* along these lines rests on *scientia* of the truths of metaphysics: in particular the truth that God exists and is no deceiver. But this form of *scientia*, once achieved, gives grounds for general confidence in methodically reached conclusions on particular scientific topics. Although Descartes thought that it was beneficial once in a lifetime to go through the metaphysical reflections that first induce and then dispel the hyperbolical doubt, he thinks that much more time should be spent on the non-metaphysical natural sciences, and he says that in his own case he spends very little time on metaphysics.

In the Cartesian natural sciences proper, hypotheses are proposed that relate questions or problems or phenomena to solutions or explanations formulated in terms of extension and motion. So long as they are drawn from a conceptual scheme serviceable for the solution of a wide array of problems, and so long as they are inferred in accordance with the precepts of Descartes's logic, they may properly be relied upon, Descartes says, even if they are not preceded or reinforced with a rehearsal of metaphysical arguments. For Descartes these hypotheses are proper objects of so-called *moral* certainty. Between the appearance of the *Discourse and Essays* (1637) and the publication of *The Principles of Philosophy* (1644), Descartes changed his mind about the strength and source of certainty of different hypotheses. He did not believe that all hypotheses could be derived from first principles, and he came to think that some deserved credence for their generality and coherence with a wide range of data they offered to explain. They deserved credence even if they sometimes posited things—such as minute particles of matter—that could not be sensed.

My own paper takes up an issue connected with moral certainty. Moral certainty about physical hypotheses is possible, according to Descartes, because *metaphysical* demonstrations show that we can trust the human faculties and methods by which we arrive at these hypotheses. Metaphysics is in that sense a fundamental science, on which physics and the rest of the derivative sciences—Descartes mentions mechanics, medicine and ethics—depend. As it is fundamental, is metaphysics *more* of a science than the derivative sciences, or is it a science in virtue of the same things as the derivative sciences: its identification of simples in a given subject matter; its solutions of problems by reasoning from simples; its complete enumerations? The answer I propose is that metaphysics is one science among others in virtue of how it relates the simple to the complex, and that it is a more basic science in virtue of the fact that its simples are more basic than—conditions for the existence of—the simples of the other sciences. This interpretation raises the question of the status in Descartes' thought of the heuristic principle of science in general that the complex depends on the simple; that problems should be resolved into questions about simples and so on. These principles belong to what Descartes calls "logic"; so perhaps *that* science deserves recognition as some sort of basic science, notwithstanding the fact that it is given no position at all in Descartes's picture of the subterranean region of the tree of science. Be that as it may, it is clear that Descartes was not out to make metaphysics into some sort of pre-eminent science, worth developing in its own right. It is more like a ladder that, kicked away, enables one confidently to propose hypotheses about nature.

If Descartes is right, there can be *scientia* about the very general features of the material world, and, perhaps even about more specific material phenomena. But what about the mental? Is this an object of *scientia*? Descartes says that the mental is better known than the physical. But does this mean that it is a more apt object of *scientia* than *matter*? Nicholas Jolley's essay suggests not. As he points out, the claim that the mind is better known than the body is prominent in Meditation Two, where what is meant is that intellection of body is at the same time acquaintance with my nature as thinking thing (AT VII 34; CSM II 22). There is no indication of an interest there in a *systematic* understanding of mental phenomena, or that this is more readily arrived at than systematic physical understanding. On the contrary, a concern with *system* is quite out of place at that stage of Descartes's metaphysical reflections: Descartes's thoughts and his confidence in them are at that point still in pieces. Anyway, Descartes does not use the term *scientia* or cognate expressions in that part of the *Meditations*. He uses *cognitio*. Things do not look significantly different either in Meditation Six, where systematic knowledge has become a possibility. Although it is a probably a matter of unshakeable certainty in Meditation Six that the mind is thinking substance, this does not open a vista on a science of the mind. The same is true, Jolley argues, in the corresponding passages of the *Principles of Philosophy*. He leaves open the question of an aspiration to *scientia* about the mind in *The Passions of the Soul*.

What Spinoza called *scientia intuitiva* owes something to *scientia* in Descartes's sense. In Descartes *scientia* begins with knowledge of God and his attributes, and knowledge of matter and mind is "deduced" from that—in the sense of being arrived at in a mental process that introduces no unclarity and that begins with a grasp of God's nature. In Spinoza, not only knowledge of matter and mind in general but the essences of particular things—finite modes—comes from knowledge of the essence of God. Don Garrett's paper shows that *scientia intuitiva* differs both from reason and from imagination and opinion. It differs from opinion and imagination in being free of distortion and confusion; it differs from the invariably clear and adequate ideas of reason, on the other hand, by always being knowledge-why and not merely knowledge-that. But though it is knowledge-why, it does not have a syllogistic medium: what follows from the essence strikes one all-at-once; it is not worked out. God's knowledge of things, both general and particular, is *scientia intuitiva*. Humanly available *scientia intuitiva* decreases the fear of death and produces the empowering joy that Spinoza called love of God. This is the characteristically Spinozistic ethical upshot of metaphysical knowledge. Descartes thought the ethical payoff of metaphysics was much more indirect.

*Scientia* in Hobbes shows traces of *scientia* in Aristotle, as Douglas Jesseph argues. Hobbes remains faithful to the Aristotelian notion that 'all *scientia* must arise from an understanding of causes'; but Hobbes's account of causes is explicitly mechanistic. The specimens of *scientia par excellence*, according to Hobbes, were, on the one hand, geometry, and, on the other, politics. This curious pairing makes sense in the light of the Hobbesian idea that the figures of geometry are constructed to have certain pre-determined properties, and that ideal polities in turn are artefacts with properties willed to exist by their makers. A demonstration of the properties of both geometrical objects and commonwealths could be understood to articulate maker's knowledge: knowledge of properties knowingly put into a thing rather than (as in physics) reconstructed as consequences of merely postulated causes. Human beings can understand *their* artefacts in rather the way that God understands his (i.e. material objects). So geometry and politics are ideal objects of human knowledge of causes, while physics is not. It is a matter of controversy whether Hobbes's theory of *scientia* aspires to be proof against scepticism, as Descartes's does, but Jesseph observes, surely correctly, that if it did aspire to this, it is a failure. Certainly Hobbes did not set out to make *physics* scepticism-proof, as he admits that there can be equally plausible but incompatible physical explanations of the same thing. But neither do his ideal sciences seem to be defended against a noticeable sceptical challenge. Instead of being scepticism-proof, scientia is intended to be the kind of knowledge that can be represented as rigorously derived from highly general principles and definitions—not "principles" and "definitions" in Aristotle's sense, but "principles" and "definitions" understood in a self-consciously unAristotelian way. So while Hobbes and Aristotle seem to agree that *scientia* results from reasoning and is systematic, it is agreement on the surface only.

Like Hobbes, Locke was an anti-Aristotelian whose doctrines sometimes betray the influence of their great adversary. Locke agreed with Aristotle that knowledge was ideally demonstrative, systematic and causal. Unlike Aristotle, he did not think that the essences of natural kinds were available through sense and induction to the human intellect. On the contrary, the human senses were simply unsuited to the acquisition of genuine causal knowledge. The deficiencies of the senses in this regard were not compensated for by an active supernatural human mind, stocked by God with ideas of simple natures needed for physics. On that score, Descartes was as much in error as Aristotle, according to Locke. Unlike ethics, which, on grounds reminiscent of Hobbes, Locke did think a demonstrative science, natural science could only be a stage for probable belief supported by experiment. Contrary to Aristotle, Locke held that we had no natural insight into the essences of natural kinds. Human science comes into its own where its starting points depend on human fiat.

The deflation of natural scientific pretension that one finds in Locke is found also in many otherwise disagreeing early modern philosophers. There is wide agreement on the point that nature and the human understanding are *not* made for one another. This consensus is deeply unAristotelian and often anti-Aristotelian. Even where the early moderns say that *scientia* in some form is possible, they tend to depart markedly from the Aristotelian picture of how it works and how available it is. They tend to depart markedly in the direction of epistemological modesty. Natural scientific demonstrations are not knock-down, not unrivalled, not a sort of natural byproduct of extended observation.

Some of the papers in this volume have been specially commissioned. Others started out as presentations to a Conference on *scientia* held in London in 2003 under the auspices of a Leverhulme Trust Academic Interchange Award (Grant F/00213A). Jill Kraye, John Rogers and I acknowledge that support with gratitude and thank the contributors for their patience.

# Contributors

**Stephen Clucas** Birkbeck College, University of London

**Daniel Garber** Princeton University

**Don Garrett** New York University

**Stephen Gaukroger** Universities of Sydney and Aberdeen

**Douglas Jesseph** University of South Florida

**Nicholas Jolley** University of California, Irvine

**G.A.J. Rogers** University of Keele

**Tom Sorell** University of Birmingham

**Catherine Wilson** University of Aberdeen

# Philosophia, Historia, Mathematica: Shifting Sands in the Disciplinary Geography of the Seventeenth Century

**Daniel Garber**

Something very important happened in our knowledge of the physical world in the seventeenth century. A number of very smart people made discoveries about the natural world that fundamentally changed our way of looking at things. But as important as the individual accomplishments of individual seventeenth-century scientists were, an important part of the story lies in the disciplinary and institutional history of that important century. What was new and important was not only Copernicus and Kepler, Descartes and Galileo, Leibniz and Newton, but the changes that happened in the larger framework in which they work. In particular, I think that there was a major change in what might be called the disciplinary geography, the way in which the disciplines that deal with our knowledge of the natural world changed in their relations with respect to one another. This involves not only intellectual changes, but, perhaps as importantly, changes in the institutions that involve the investigation of nature and the dissemination of new knowledge and points of view.

The story is very big and very complex, and I cannot hope to tell it all in a single article. But in this short essay I would like to give a bare outline of what the whole story might look like. I will begin with an overview the state of the disciplines that are concerned with the natural world ca. 1600. I will then trace through the way in which the disciplinary geography changes over the course of the century.

## Scientia, Mathematica and Historia ca. 1600

In the recent literature, there has been considerable debate about the scientific revolution of the seventeenth century. Steve Shapin began his little book, *The Scientific Revolution* with the following provocative statement: "There was no such thing as the Scientific Revolution, and this is a book about it." One reason for Shapin's skepticism is the fact that in the seventeenth century, there was no such thing as science, as we understand the term:

---

D. Garber (✉)
Princeton University, Princeton, NJ, USA

T. Sorell et al. (eds.), *Scientia in Early Modern Philosophy*, Studies in History and Philosophy of Science 24, DOI 10.1007/978-90-481-3077-1_1,

> Many historians are now no longer satisfied that there was any singular and discrete event, localized in time and space, that can be pointed to as "the" Scientific Revolution. Such historians now reject even the notion that there was any single coherent cultural entity called "science" in the seventeenth century to undergo revolutionary change. There was, rather, a diverse array of cultural practices aimed at understanding, explaining, and controlling the natural world, each with different characteristics and each experiencing different modes of change.[1]

I think that this is basically right, at least in the beginning of the century.

When I say that there was no such thing as science in 1600, I certainly don't mean to say that people weren't interested in the things that we now call scientific, in stars and plants, in falling balls and magnets and so many other things. They were, of course, just as people had been interested in nature from virtually as far back as we have information about people and their interests. My point is just that these things didn't constitute a kind of intellectual natural kind, or at least, the intellectual natural kinds didn't correspond exactly to what they do now, what we call science.

When approaching the history of science in the seventeenth century, it is important to realize that the word "science" simply does not mean what it does for us. "Scientia", the Latin cognate, meant simply "knowledge", albeit knowledge in a rather strict sense that entailed certainty. But this is not just a linguistic point: there is no intellectual category at the beginning of the seventeenth century that exactly corresponds to what we call science.

To understand this point, let us begin with the intellectual geography outlined in Francis Bacon's important and influential *Advancement of Learning* (1605), published at the very beginning of this fecund century. Bacon categorizes knowledge as follows:

> The parts of human learning have reference to the three parts of man's understanding, which is the seat of learning: history to his memory, poesy to his imagination, and philosophy to his reason. (Bacon, *Advancement*, II I.1)[2]

Of these, history and philosophy are most relevant to our interests here. In Bacon's later Latin expansion of the *Advancement of Learning*, his *De dignitate et augmentis scientiarum* (1623) history is characterized as follows:

> History is properly concerned with individuals, which are circumscribed by place and time. For though Natural History may seem to deal with species, yet this is only because of the general resemblance which in most cases natural objects of the same species bear to one another; so that when you know one, you know all. And if individuals are found, which are either unique in their species, like the sun and moon; or notable deviations from their species, like monsters; the description of these has as fit a place in Natural History as that of remarkable men has in Civil History. (Bacon, *De dignitate*, II.1)

Goclenius offers a similar definition in his authoritative *Lexicon philosophicum* of 1613:

---

[1] Steven Shapin, *The Scientific Revolution* (Chicago: University of Chicago Press, 1996), pp. 3–4.

[2] References to Bacon's *Advancement of Learning* and *De dignitate et augmentis scientarum* are given in the text in a way that should allow the reader to find them in any standard edition.

> History means the awareness of singular things, either setting them out or describing them. . . . [History] is understood by philosophers . . . as observation, and as knowledge from one's own experience, that is, as belief derived from the senses over the course of time.[3]

History, then, deals with particulars.

Bacon recognizes a number of different kinds of history, including civil, ecclesiastical, and literary, in addition to natural history. Natural history itself is divided into a number of different sorts. Bacon writes:

> History of nature is of three sorts; of nature in course, of nature erring or varying, and of nature altered or wrought; that is, history of creatures, history of marvels, and history of arts. (Bacon, *Advancement*, II 1.3)

The history of "nature in course" is simply the observational study of things as normally found in nature. This is natural history as we have come to understand it in the normal sense, the study of plants and animals in nature, of heavenly bodies, geological formations, etc. The history of "nature erring or varying" is the study of monsters and marvels, the two-headed calves and the like. The history of "nature altered or wrought" is the study of nature constrained in various way. This includes the study of machines, devices built by art for particular human purposes, as well as the experimental study of nature.

For Bacon, philosophy is something quite distinct from history. Bacon divides philosophy into three main parts, divine philosophy, concerned with God, natural philosophy, concerned with nature, and "human philosophy or humanity," concerned specifically with man. The "common parent" of all three is what Bacon calls "philosophia prima," something that he distinguishes from metaphysics. (Bacon, *Advancement*, II 5; 7.3) Natural philosophy, what is of most concern to us here, is divided into two parts, "the inquisition of causes, and the production of effects; speculative, and operable; natural science and natural prudence." (Bacon, *Advancement*, II 7.1) The operative side of natural philosophy is what he calls natural magic, properly understood, "that great liberty and latitude of operation which dependeth upon the knowledge of Forms." (Bacon, *Advancement*, II 8.3) Theoretical natural philosophy, the search after causes, divides into two parts, physics and metaphysics:

> For as we divided natural philosophy in general into the inquiry of causes, and productions of sects: so that part which concerneth the inquiry of causes we do subdivide according to the received and found division of causes; the one part, which is Physic, inquireth and handleth the material and efficient causes; and the other, which is Metaphysic, handleth the formal and final causes. (Bacon, *Advancement*, II 7.3)

And so Bacon summarizes his categorization of the branches of knowledge that deal with nature:

> Physic, taking it according to the derivation, and not according to our idiom for medicine, is situate in a middle term or distance between Natural History and Metaphysic. For natural

---

[3] Rudolph Goclenius, *Lexicon philosophicum, quo tanquam clave philosophiae fores aperiuntur* (Frankfurt: 1613), p. 626.

> history describeth the variety of things; physic, the causes, but variable or respective causes; and metaphysic, the fixed and constant causes. (Bacon, *Advancement*, II 7.4)

As he full well knew, Bacon is using the term "metaphysics" in a rather idiosyncratic way here. But his conception of physics or natural philosophy as dealing with the underlying causes in nature is quite orthodox. As Eustachius a Sancto Paolo put it in his very popular textbook in philosophy, first published in 1609, "The natural philosopher explains many effects of nature through their true causes."[4] Natural philosophy, otherwise known under the names of physics or physiology, dealt first with the general truths about natural things (matter, form, privation, cause, etc. for the Aristotelian, e.g.), then with the more specific truths about more specific things, including the cosmos as a whole, the elements, the make up of individual kinds of bodies, including all kinds of living bodies, plant, animal, and human.

"Natural philosophy" or "physics" in this technical sense is probably the closest term Bacon and his contemporaries had at the beginning of the century to the modern term "science". But the match is only partial. As noted earlier, natural philosophy is an enterprise quite distinct from natural history, another systematic way of approaching the study of nature. But there is another enterprise related to the study of nature, mathematics. In the *Advancement of Learning*, Bacon includes mathematics as a part of natural philosophy, and numbers it rather awkwardly as part of metaphysics. A particularly important part of mathematics is what Bacon and his contemporaries called mixed mathematics. Bacon writes:

> The Mathematics are either pure or mixed. To the Pure Mathematics are those sciences belonging which handle quantity determinate, merely severed from any axioms of natural philosophy; and these are two, Geometry and Arithmetic; the one handling quantity continued, and the other dissevered. Mixed hath for subject some axioms or parts of natural philosophy, and considereth quantity determined, as it is auxiliary and incident unto them. For many parts of nature can neither be invented with sufficient subtlety, nor demonstrated with sufficient perspicuity, nor accommodated unto use with sufficient dexterity, without the aid and intervening of the mathematics; of which sort are perspective, music, astronomy, cosmography, architecture, enginery, and divers others. (Bacon, *Advancement*, II 8.2)

Connected with the mathematical sciences, particularly "enginery" or mechanics, is one of the branches of natural history mentioned above, what Bacon called the history of nature altered or wrought, the history of the arts, or *historia mechanica*. About this Bacon writes as follows:

> But if my judgment be of any weight, the use of history mechanical is of all others the most radical and fundamental towards natural philosophy; such natural philosophy as shall not vanish in the fume of subtile, sublime, or delectable speculation, but such as shall be operative to the endowment and benefit of man's life: for it will not only minister and suggest for the present many ingenious practices in all trades, by a connection and transferring of the observations of one art to the use of another, when the experiences of several mysteries shall fall under the consideration of one man's mind; but further, it will give a more true and

---

[4]Eustachius a Sancto Paolo, *Summa philosophiae quadripartita, de rebus Dialecticis, Ethicis, Physicis, & Metaphysicis* (Paris: 1609). The quotation is from the edition published in Cambridge, 1648, *Physica*, p. 111.

> real illumination concerning causes and axioms than is hitherto attained. For like as a man's disposition is never well known till he be crossed, nor Proteus ever changed shapes till he was straitened and held fast; so the passages and variations of nature cannot appear so fully in the liberty of nature, as in the trials and vexations of art. (Bacon, *Advancement*, II 8.2)

But, Bacon notes, there is a deep prejudice against inquiring into this kind of history:

> For history of nature wrought or mechanical, I find some collections made of agriculture, and likewise of manual arts; but commonly with a rejection of experiments familiar and vulgar. For it is esteemed a kind of dishonour unto learning to descend to inquiry or meditation upon matters mechanical, except they be such as may be thought secrets, rarities, and special subtilities... (Bacon, *Advancement*, II 8.2)

More generally, it is fair to say that practitioners of the higher sciences, natural philosophy, physics considered themselves quite distinct from the mechanics and other mixed mathematicians, those who dealt with mere practical matters. Though Bacon included mathematics, mixed and pure within natural philosophy, most of his contemporaries would have seen it as something quite distinct. An interesting symbolic representation of the relation between mixed mathematics and physics is given in the frontispiece to Niccolò Tartaglia's important *Nova Scientia* (Venice: 1537)(See Fig. 1). In that image, there are two circular walls, one higher than the other. At the entrance to the larger enclosure, Euclid stands guard, making sure that that no one enters who does not know geometry. In the first level, the larger of the two, stands Tartaglia himself, surrounded by figures standing for Arithmetic and Geometry, together with representatives of the various mathematical sciences, including astronomy, astrology, and music. This level also contains representations of the cannons and cannon balls that form the major subject matter of the book. The higher enclosure is labeled as the domain of philosophy. Guarded by Plato and Aristotle, it is obviously a higher calling, more difficult to enter; in Tartaglia's engraving, it stands empty, except for Philosophia herself. In the diagram, it is clear that the two disciplines are distinct: philosophy, which, presumably includes natural philosophy, physics, is a discipline different from the mathematical sciences. Bacon's call to see the connections between the mechanics and the physicists, between theory and practice were quite radical: they were a call to a new philosophy rather than observations on current intellectual practice.

Also distinct from natural philosophy for Bacon were the sciences that concern human beings, both their minds and bodies. Though obviously connected with natural philosophy, Bacon separates these sciences, which, properly speaking, ground medicine, from the other parts of natural philosophy and gives them extensive separate treatment. Again, Bacon attempts to relate the sciences of man to the other parts of natural philosophy. But, again, in Bacon this is a program rather than a reflection of current practice.

In these different domains of inquiry we can see much of what now goes on under the heading of science. But it is important to note that despite the common threads that we can see connecting these different domains, in 1600 they did not come together as a unified project, what we would call science in the modern sense,

**Fig. 1** Frontispiece to Niccolò Tartaglia, *Nova scientia* (1537)

a general domain of inquiry into nature: they are a collection of disparate projects. While they did have connections with one another, and while Bacon, for one, hoped that they would become more connected, there was no single term or concept that united them all. This is further underscored when we examine their larger context in Bacon's organization of knowledge. In presenting Bacon's views, I have pulled out the parts of the *Advancement of Learning* that look most like modern science. But this took a certain amount of reworking. In Bacon's text it is important to note that natural history is linked not with natural philosophy, but with the other kinds of history, civil, ecclesiastical, and literary; natural philosophy itself is linked with other kinds of science, including divine and human; the science of man includes not only the study of mind and the anatomy of the body, but also politics. There are some familiar landmarks in Bacon's map of knowledge, but the larger geography is very different from ours.

## People and Places

In the previous section we outlined some of the different intellectual categories into which knowledge fit in 1600. These distinctions are not merely of intellectual importance: in many ways they mirror real differences between practitioners of the different disciplines and the contexts in which they practiced them.

Perhaps the easiest place to begin is with natural philosophy. Natural philosophy was a part of the studies undertaken by students under the direction of the Faculty of Arts, part of their preparation for entering into the advanced faculties of medicine, law, or theology. Typical here is the course of studies outlined in the *Ratio studiorum* of the Jesuit order, a plan of studies, finalized in 1599, that set out how all Jesuit schools were to be organized.[5] According to the *Ratio studiorum*, the course in philosophy was to take three years, and be taught for 2 hours per day, one before noon, the other after noon. The entire first year was to be taken up with logic. In the second year, students were to study Aristotle's *Physics*, *De caelo*, and the beginning of the *De generatione et corruptione*. In the third year, students were to be taught the rest of the *De generatione et corruptione* and the *De anima*, as well as the *Metaphysics*. The *Meterologica* was to be taught in the summer between the second and third years of the philosophy course. Altogether, then, natural philosophy would take up about a year and a half, about half the time spent on philosophy as a whole. The teachers of such courses were generally not what we would call philosophers, scientists, or physicists: they were teachers whose role it was to transmit an established body of doctrine to a group of adolescents. Typically, such teachers would remain in the arts faculty for 5 or 10 years at the most, before moving to one of the higher faculties or outside the university altogether, as preachers, missionaries, or administrators in their orders. In general it is fair to say that in the early seventeenth

[5] See *Ratio studiorum: Plan raisonée et institution des études dans la Compagnie de Jésus*, ed. and trans. by Adrien Demoustier et al. (Paris: Belin, 1997). This contains the *Ratio studiorum* of 1599, in Latin with French translation, notes and supplementary material.

century the colleges and universities were not sites of active original research in natural philosophy.[6]

Active discussion and research into natural philosophy took place outside the universities. But at the beginning of the century, there is no single place where such activities took place. The young Galileo taught Aristotelian physics as a university professor; the manuscripts that he left include lecture notes for courses that he likely taught on that subject, in addition to the mathematical subject that he also taught. But as soon as he could, he escaped the university and became attached to a court, as the Philosopher and Mathematician to the Grand Duke of Tuscany.[7] William Gilbert, the author of the important *De magnete* (1600), was a London physician. Descartes, whose mechanical philosophy challenged the Aristotelian natural philosophy had a small personal income, and worked in isolation in Holland. Though his ideas were introduced into the Dutch universities during his lifetime, that can hardly be considered a success. They caused law suits and nothing but trouble for the unfortunate French natural philosopher.[8]

In the beginning of the seventeenth century, mathematics was taught in the colleges and universities, alongside natural philosophy.[9] However, it is important to note here that as it was taught in the colleges and universities, it was generally taught by a teacher different from the teacher who taught natural philosophy, along side of it, but not equal. In the Jesuit colleges, for example, while natural philosophy would receive at least 2 hours of instruction every day in the second and first half of the third year of instruction in philosophy, mathematics would get only 45 minutes. Including even that much was not easy; conservative Aristotelian philosophers within the Jesuit order and its main academic institution, the Collegio Romano in Rome, opposed even that. It was only due to the assiduous pressure of the Jesuit mathematician Christopher Clavius that even that much was included.[10]

The separation between the two domains in Tartaglia's imaginative representation discussed earlier seems to have corresponded to practice. Recent studies have emphasized the social differences between the practitioners of mathematics and that of natural philosophy in the years leading up to the seventeenth century. Mathematics was generally considered the more practical discipline, closely associated with

---

[6]For a good account of the organization and structure of universities in France in the early seventeenth century, see L.W.B. Brockliss, *French Higher Education in the Seventeenth and Eighteenth Centuries: A Cultural History* (Oxford: Oxford University Press, 1987)

[7]See Stillman Drake, *Galileo at Work* (Chicago: University of Chicago Press, 1978) and William Wallace, *Galileo and his Sources: The Heritage of the Collegio Romano in Galileo's Science* (Princeton: Princeton University Press, 1984)

[8]See Theo Verbeek, *Descartes and the Dutch: Early Reactions to Cartesianism (1637–1650)* (Carbondale, IL: Southern Illinois University Press, 1992).

[9]See Mordechai Feingold, *The Mathematicians' Apprenticeship: Science, Universities, and Society in England, 1560–1640* (Cambridge: Cambridge University Press, 1984); François de Dainville, *L'education des jésuites (XVIe–XVIIIe siècles)* (Paris: Les Editions de Minuit, 1978), pp. 323–54; *Ratio studiorum*, p. 132.

[10]See Giuseppe Cosentino, "L'insegnamento delle matematiche nei collegi gesuitici nell'Italia settentrionale: Nota introduttiva," *Physis* 13 (1971), pp. 205–17.

practical mechanics, military architecture and the like, and its practitioners were often of lower social standing than those who practiced natural philosophy. In the universities and in the courts, it was the natural philosophers who got paid more and who had the higher social status. One can read Galileo's career as in good part an attempt to boost the status of the mathematician and give him a status equal to that of the philosopher. This will change as the century goes on, but at the beginning of the seventeenth century, the two communities were quite distinct.[11]

But there were more communities still who were interested in nature from other points of view. Another community of savants interested in nature were the alchemists. While Bacon left little place for them in his catalogue of knowledge, alchemy and alchemists were an extremely important part of the pre-scientific community of the early seventeenth century. The idea of alchemy meant many things to many people in the period, and it is very dangerous to generalize. Alchemy was both theory and practice, involving both an account of at least a part of the natural world, and an application of that understanding to the practical problems of transforming base metals into gold and silver, as well as other aspects of what we might now call chemical engineering, as well as the problem of curing patients. For some, the theoretical part of alchemy dealt with only a part of nature, with mixtures or with metals. But for some, alchemy was itself the whole of natural science, a genuine natural philosophy, and a conception of the foundations of natural science alternative to that offered by the Aristotelians insofar as chemical philosophers offered an alternative conception of the basic categories and principles of the physical world. Alchemists occupied a wide range of roles in society. Some were in the University, particularly in the faculties of medicine; some were at courts, particularly in the German-speaking countries. Many were practitioners of alchemy as a trade, either connected with medicine, or with metallurgy and the like. In this capacity they were often threatened with legal action when they failed to produce the promised results, or when they strayed from their spiritual calling and became counterfeiters.[12]

Also important as students of nature were the medical community. All physicians had passed through the faculty of arts, of course, and had been trained in Aristotelian natural philosophy. But after that the path diverged. From the arts faculty students passed to the medical faculty, where they read Hippocrates and Galen, or, in some places, Paracelsus and his followers, and were trained in the medical arts. Distinct from the physicians were others who cared for the body, including

[11] See Robert S. Westman, "The Astronomer's Role in the Sixteenth Century: A Preliminary Study," *History of Science* 18 (1980), pp. 105–47; Mario Biagioli, "The Social Status of Italian Mathematicians, 1450–1600," *History of Science* 27 (1989), pp. 41–95; Mario Biagioli, *Galileo Courtier: The Practice of Science in the Culture of Absolutism.* (Chicago: University of Chicago Press, 1993).

[12] On alchemy (or "chymistry" as some current historians prefer) see especially Pamela Smith, *The Business of Alchemy: Science and Culture in the Holy Roman Empire* (Princeton, NJ: Princeton University Press, 1994); William Newman, *Gehennical Fire*, (Cambridge, MA: Harvard University Press, 1994); Lawrence M. Principe, *The Aspiring Adept: Robert Boyle and His Alchemical Quest* (Princeton: Princeton University Press, 1998).

surgeons and apothecaries, who had their own guilds and their own distinct patterns of training. They also had their own special places of research. While some worked in the medical faculties of universities, many worked in courts, hospitals, clinics, and in botanical gardens.[13]

And finally, there were the natural historians. Bacon wanted to integrate natural history and natural philosophy; natural history was to be the foundation of his new natural philosophy. But at the beginning of the century, at least, the two were quite separate. In general, natural history had little connection with the university in the early seventeenth century. It also had little connection with anyone involved with natural philosophy, either the older Aristotelian natural philosophy, or the newer varieties of natural philosophy that were coming to challenge the Aristotelian natural philosophy. Natural history was most closely connected with the voyages of discovery that were bringing new plants, animals and minerals back to Europe, and to the cabinets of curiosity and wonder that were formed to exhibit these new objects of wonder. They were also associated with the botanical and zoological gardens formed to study these new specimens.[14]

My point here is that there is no common group of people in 1600 that one can call scientists, and no common institutions that they share. Natural philosophy was most often the province of the Arts Faculty of the universities, less a research subject (certainly not an experimental research subject) than a part of the teaching curriculum for students in their teens, before they went on to the higher faculties of the University. Medicine (and the sciences relating to the human body, including anatomy) would be taught by the medical faculty of the University, something quite separate from the Arts faculty. The medical faculty might also deal with alchemy, insofar as it relates to medicines, with botany (insofar as it relates to materia medica), and even to astrology. The mathematical arts could be carried out in a number of places: at the university, as a supplement to (but quite distinct from) the curriculum in natural philosophy, at the court or the arsenal, insofar as it related to more practical matters, both military and non-military, on the work-site. What we call biology could be carried out at the various gardens, founded sometimes for medical research, sometimes for biological research, sometimes simply out of curiosity, to display the wealth of imperial conquest. Sometimes the private collections and cabinets of curiosity of private collectors served as research tools for biological research. These different enterprises were not always entirely unrelated, of course. But they were often different and distinct: there was no one natural kind, be it "science" or "natural philosophy" under which they all fell, and their practitioners were all quite different, ranging from university professors to doctors to alchemists in business for themselves to gentleman collectors.

---

[13] See L.W.B. Brockliss and Colin Jones, *The Medical World of Early Modern France* (Oxford: Oxford University Press, 1997).

[14] See Krzysztof Pomian, *Collectors and Curiosities* (Cambridge: Polity Press, 1990); Paula Findlen, *Possessing Nature: Museums, Collecting and Scientific Culture in Early Modern Italy* (Berkeley and Los Angeles: University of California Press, 1994); N. Jardine, J.A. Secord, and E.C. Spary, eds., *The Cultures of Natural History* (Cambridge: Cambridge University Press, 1996).

## What Happened in the Seventeenth Century?

So far I have been emphasizing the ways in which the domain of knowledge differed from our own at the beginning of the seventeenth century. But there were major changes over the course of the century.

There is no question about it: the seventeenth century was a period remarkable for its scientific fecundity, something that was recognized even by those who lived through it. Figures such as Bacon and Descartes certainly recognized their own outstanding intellectual qualities, and were eager to tell their contemporaries about the remarkable new start that they were providing for the sciences, a new and deeper understanding of nature that broke sharply with that of past generations. Descartes, for example, after a few modest words about his own limitations, immodestly confesses to have found a new natural philosophy that gives humankind the key to nature itself:

> But as soon as I had acquired some general notions in physics and had noticed, as I began to test them in various particular problems, where they could lead and how much they differ from the principles used up to now, I believed that I could not keep them secret without sinning gravely against the law which obliges us to do all in our power to secure the general welfare of mankind. For they opened my eyes to the possibility of gaining knowledge which would be very useful in life, and of discovering a practical philosophy which might replace the speculative philosophy taught in the schools. Through this philosophy we could know the power and action of fire, water, air, the stars, the heavens and all the other bodies in our environment, as distinctly as we know the various crafts of our artisans; and we could use this knowledge – as the artisans use theirs – for all the purposes for which it is appropriate, and thus make ourselves, as it were, the lords and masters of nature.[15]

Bacon is even more concrete than Descartes about what we can expect from the new science he brings us. In the *New Atlantis*, Bacon's science fiction story about a society grounded in his conception of science, he goes into exquisite detail about the numerous discoveries that the members of the society's research institute, the House of Saloman are supposed to have made, and the numerous technological marvels that they are supposed to have invented, including what we can now recognize as the telephone, the television, airplanes, genetic engineering of plants and animals, miracle drugs, and on, and on, and on.[16] But much more significant, perhaps, is the fact that others saw the remarkable accomplishments of the new science as well. In 1668 the English poet John Dryden wrote:

> Is it not evident, in these last hindered years (when the Study of Philosophy has been the business of all the *Virtuosi* in *Christendome*) that almost a new Nature has been reveal'd to us? that more errours of the School have been detected, more useful Experiments in Philosophy have been made, more Noble Secrets in Opticks, Medicine, Anatomy, Astronomy,

---

[15]Descartes, *Discours de la méthode*, in Descartes, ed. Charles Adam and Paul Tannery, *Oeuvres de Descartes* (11 vols.) (Paris: CNRS/Vrin, 1964–1974), vol. VI, pp. 61–62, translated in Descartes, ed. and tr. by John Cottingham et al. *The Philosophical Writings of Descartes* (3 vols.) (Cambridge: Cambridge University Press, 1984–1991), vol. I pp. 142–3.

[16]See Bacon, *New Atlantis* (London: 1627), pp. 31 ff.

discover'd, than in all these credulous and doting Ages from *Aristotle* to us? so true it is that nothing spreads more fast than Science, when rightly and generally cultivated.[17]

The progress represented by the new science of the seventeenth century was a standard trope in the Enlightenment of the eighteenth century. This comes through loud and clear in the *Encyclopédie*. In the article entitled "expérimental", for example, d'Alembert tells the story of the birth of the new science, the experimental philosophy, as he calls it. Beginning with Bacon and Descartes and their rejection of Aristotle and his commentators, d'Alembert takes the story through the Italian académie del Cimento, to Boyle and Mariotte, until he finally arrives at Newton. Newton, according to d'Alembert, showed us "how to introduce geometry into physics, and to shape it by reducing experiment to calculation, a science exact, profound, luminous and new." He continues: "A new generation has arisen, since when the foundations of a revolution are once thrown down, almost always it is in the following generation that the revolution is brought to a conclusion."[18] The story is basically the same in d'Alembert's *Discours préliminaire* to the *Encyclopédie*, but in addition to mentioning Bacon, Descartes, Boyle, Mariotte, and Newton, he adds comments about Galileo, Harvey, Huygens, Malebranche, Leibniz, Sydenham, Boerhaave, as well as "une infinité d'Anatomistes et de Physiciens célébres."[19]

This is not the place to rehearse the numerous important discoveries and innovations that took place during the seventeenth century. But it is also important to note another subtler kind of change that one finds in the period, a reorientation and unification of the disparate disciplines concerned with the study of the natural world. I mentioned briefly the way in which mixed mathematics came more and more to take over the role earlier played by physics or natural philosophy in the intellectual sphere. In the beginning of the century, mathematics and natural philosophy (physics) were separate domains. When Newton published his great work in 1687, he entitled it *Mathematical Principles of Natural Philosophy*, firmly linking mathematics and natural philosophy. In 1600 this title would have been virtually unintelligible; by 1687 this was a title that one could almost take for granted. During this period one also saw the linking of *scientia* and *historia*, natural philosophy and natural history, the other great dichotomy to which I had referred. Here the key figure was Sir Francis Bacon. A key work here was Bacon's very influential *Novum organum* (1620), his outline of a new method for finding unlocking nature's secrets. Bacon's main interest in that work was a program in natural philosophy, finding the underlying forms of things, that which will explain how it is that sensible qualities are produced. And so, for example, Bacon shows in part II of the *Novum organum* how we can find the underlying form of heat, which turns out to be a variety of motion. But the method that Bacon presents is grounded in

---

[17]Quoted in H. Floris Cohen, *The Scientific Revolution: A Historiographical Inquiry* (Chicago: University of Chicago Press, 1994), p. 1.

[18]Diderot, Denis; Alembert, Jean Le Rond d', et al., *Encyclopédie, ou Dictionnaire raisonné des sciences, des arts et métiers* (Paris: 1751–1772), vol. 6, p. 299.

[19]Diderot et al., *Encyclopédie*, vol 1, p. xxviij.

the observation of particulars. Bacon shows us how to make up tables of particular observations of various sorts. It is through a systematic examination of the observations made on particulars, natural history, that we find the general truths of natural philosophy that Bacon is seeking. In this way it is through natural *history* that we ultimately attain natural *philosophy*. Just as Descartes, Hobbes and Newton linked mathematics with physics, Bacon shows us how to link natural history with physics. The program Bacon started is also reflected in Locke's *Essay Concerning Humane Understanding*. He writes:

> This way of getting and improving our knowledge in substances only by experience and history, which is all that the weakness of our faculties in this state of mediocrity, which we are in this world, can attain to; makes me suspect, that natural philosophy is not capable of being made a science. We are able, I imagine, to reach very little general knowledge concerning the species of bodies, and their several properties. Experiments and historical observations we may have, from which we may draw advantages of ease and health, and thereby increase our stock of conveniences for this life; but beyond this I fear our talents reach not, nor are our faculties, as I guess, able to advance.[20]

In this way Locke rejects the idea that we can have a genuine science of body in the classical sense, a deductive science, certain and grounded in first principles. Instead he recommends a Baconian natural philosophy, one grounded in a natural history, the knowledge of particulars. Once again, two categories that were once quite distinct from one another come together. By the end of the seventeenth century, the domains of natural philosophy, mixed mathematics, and natural history are merging into one enterprise.

But the unification of the domain of science is not merely a question of these particular intellectual transformations, the linking of natural philosophy with mathematics and natural history. Connected with these intellectual transformations there were was an extremely important institutional transformation: the creation of academies and journals that could serve as a unified site for the investigation of nature in different ways.

As we discussed earlier in this essay, in the early seventeenth century, there wasn't a single place where what we would call science was done. The University was not a research institute, as it is now: it was a place where you taught students. In France, at the Collège Royal there was somewhat more variety. By the late 1640s, it came to be a home for mathematicians (largely *mixed* mathematicians) such as Roberval and Gassendi, as well as a certain number of astronomers.[21] There were also the informal and private academies. In Italy there was the academie del Cimento, a private experimental academy.[22] In Paris there was also Mersenne's private academy, a collection of friends, including Hobbes, Roberval, Gassendi, the

[20] Locke, *Essay Concerning Human Understanding*, 4.12.10.

[21] See Claude-Pierre Goujet, *Mémoire Historique et Littéraire sur le Collège Royal de France* (Paris: 1758); and Abel Lefranc, *Histoire du Collège de France* (Paris: Hachette, 1893).

[22] See, e.g., W.E. Knowles Middleton, *The Experimenters: A Study of the Accademia del Cimento* (Baltimore: Johns Hopkins University Press, 1971); Paolo Galluzzi, "L'Accademia del Cimento: "Gusti" de principe, filosofia e ideologia dell'esperimento," *Quaterni storici* 16 (1981), 788–844.

Pascals, and others, who met regularly in his chambers in the Minim convent near the Place Royal to discuss new ideas.[23] The Jardin du roi in Paris served as a meeting place for alchemists to talk.[24] But there was nothing at that point that one could call a scientific institution.

However, in 1627, a year after his death, Bacon's friends published his *New Atlantis*. In the *New Atlantis*, Bacon outlined the idea of a brand new kind of institution. Bacon called it Salomon's House. It was an organization, supported by the government, for the purpose of conducting a variety of investigations into nature. Organized in a way that reflected Bacon's conception of investigation in the *Novum organum*, it included members that gathered observations, performed experiments, drew theoretical conclusions from those experiments, and translated the theoretical knowledge into practical benefits for society. As one member of Salomon's House put it, "The end of our foundation is the knowledge of causes, and secret motions of things; and the enlarging of the bounds of human empire, to the effecting of all things possible."[25] Included in purview of Salomon's House is everything that would later be called scientific, from natural history to biology to anatomy and medicine to mechanics and physics, united under one roof, and thereby given a corporate identity. Salomon's House would later be built, in London as the Royal Society (founded 1660) and in Paris as the *Académie Royale des Sciences* (founded in 1666), societies that were imitated all over Europe, both in large cities and in smaller provincial cities who wanted to enhance their status.

The *Académie Royale des Sciences* housed some of the most distinguished mathematicians, mixed and pure, of the day, including Huygens, Carcavi, and Cassini, as well as distinguished natural historians such as Claude Perrault. The Paris *Académie* was obviously involved in discussions of mathematical subjects such as Cassini's astronomy and Huygens' work on motion and mechanics at the highest level. But, at the same time, it also published the monumental *Mémoires pour servir à l'histoire naturelle des animaux* (1671–1676) and the *Mémoires pour servir à l'histoire naturelle des plantes* (1676). In the *Académie des sciences*, a spirit of cooperation reigned to the point that the astronomers even attended and commented on the anatomical dissections.[26]

The Royal Society of London had a broader membership than did the *Académie Royale des sciences*, more nobility and therefore more in the way of amateur virtuosi, and fewer in the way of what were emerging as scientific professionals.

---

[23] See, e.g., J.-R. Armogathe, "Le groupe de Mersenne et la vie académique parisienne," *XVIIe Siècle* 1992, 44:131–39.

[24] See Dr. E.T.H. Hamy, "William Davisson, intendant des jardins du roi et professeur de chimie," in *Nouvelles Archives du Museum* 3e S., t. X, Paris, 1898, pp. 1–38; E.C. Spary, *Utopia's Garden: French Natural History from the Old Regime to Revolution* (Chicago: University of Chicago Press, 2000).

[25] Francis Bacon, ed. J. Spedding, R. L. Ellis, D. D. Heath, *The Works of Francis Bacon*, (14 vols.) (London: Longmans, 1857–1874), vol. III, p. 156.

[26] See Roger Hahn, *The anatomy of a scientific institution: the Paris Academy of Sciences, 1666–1803* (Berkeley and Los Angeles: University of California Press, 1971).

However, its interests were just as varied as the Paris group. In 1665 they published Robert Hooke's monumental *Micrographia*, the record of his microscopical investigations. But in 1687 they also sponsored the publication of Newton's *Principia*, a work that defined mathematical physics.[27] The variety of their interests can also be seen in their house journal, the *Philosophical Transactions of the Royal Society*. From the beginning, the *Philosophical Transactions* included a wide variety of different kinds of materials. The first issues contained, for example, astronomical observations, the reports of optical experiments, and other bits of mathematical science. But they also contained a review of Hooke's *Micrographia*, a review of Boyle's *Natural History of Cold*, observations about monstrous births, and other bits of natural history. This mix of articles is characteristic of what will appear in the pages of the *Philosophical Transactions* for the rest of the century.

The *Philosophical Transactions* represented a radically new way of communicating ideas. In 1600, circulating new ideas was not easy. It was, of course, always possible to publish a book, if one had the money to do so, or if one could convince a bookseller that it would be worth his while. Alternatively, one could publish a pamphlet, or a broadside, and post it in a public place. In Paris, for example, the Pont Neuf was a very popular place to distribute ideas in that way. Much communication went by way of letter. The letter was not always the private kind of communication it is today. Mersenne was the master of this kind of communication. He might send a letter to Descartes, say, with questions gleaned from his circle. The reply would then be copied and distributed to various members of his circle, who might, in turn, reply to what Descartes had written. Collections of these letters might, of course, make it into print. Sometimes Mersenne would take particularly interesting bits and publish them in some of his own books, such as the *Cogitata physico-mathematica* (Paris, 1644), where Mersenne reports on the ideas of Descartes, Hobbes, and Roberval, for example. But in the 1660s, there arose a new form of communication, the journal. The *Philosophical Transactions* first appeared in 1665, and in the same year, in Paris appeared another journal, the *Journal des sçavans*. In 1682 first appeared the *Acta Eruditorum* in Leipzig, associated with Leibniz; this was to be the journal in which Leibniz published the first notices of his new calculus. This was followed in 1684 by the *Nouvelles de la république des lettres*. With the exception of the *Philosophical Transactions*, none of these are scientific journals in the narrow sense that we now understand the term. In addition to mathematical and natural historical material, they included communications on theology, history, current events, literature. But in the pages of all of these publications appeared some of the seminal articulations of what the eighteenth century would see as the new science.

In this way, by the end of the seventeenth century and the beginning of the eighteenth, we can say that there has emerged a new institution, a community of practitioners, independent of the University and organized religion, a group joined by

---

[27] See Michael Hunter, *The Royal Society and its Fellows, 1660–1700: the Morphology of an Early Scientific Institution* (Chalfont St. Giles, Bucks, England: British **Society** for the History of Science, 1982); Michael Hunter, *Establishing the New Science: The Experience of the Early Royal Society* (Woodbridge, Suffolk; Wolfeboro, NH, USA: Boydell Press, 1989).

journals and societies, bringing together under one roof, into one community what had been earlier in the century a variety of different practitioners, natural philosophers, mathematicians, natural historians, giving them a home and, most importantly, an identity. In this way the different disciplines concerned with nature came together to form what increasingly looked like a single domain of inquiry, what would later be called "science".

Coordinate with the creation of science as an enterprise, both intellectually and institutionally, was the increasing distinction between what would be called philosophy in the modern sense, and what would be called the sciences. In 1600, "philosophia" and "scientia" were virtually synonymous; both referred to knowledge in the strong sense, certain, general, and grounded in the knowledge of causes. But by the end of the century there is an increasing distinction between the empirical and mathematical study of nature, and another way of contributing to knowledge. In a famous passage, Locke writes:

> The commonwealth of learning is not at this time without master-builders, whose mighty designs in advancing the sciences, will leave lasting monuments to the admiration of posterity; but every one must not hope to be a Boyle, or a Sydenham; and in an age that produces such masters as the great Huygenius, and the incomparable Mr. Newton, with some others of that strain; it is ambition enough to be employed as an under-labourer in clearing the ground a little, and removing some of the rubbish that lies in the way to knowledge... (Locke, *Essay*, Epistle to the Reader)

Now, it is not entirely clear how Locke himself meant this passage. But it is clear that later eighteenth-century readers of this widely-read book saw in it a distinction between what we would now call philosophy and what we would now call science. It is interesting how d'Alembert characterizes Locke and Newton in his *Discours préliminaire* to the *Encyclopédie*. As I noted earlier, for d'Alembert, Newton is the culmination of the story about the scientific revolution, the one who "gave Philosophy the form which, it appears, it ought to keep." But, d'Alembert reports, while Newton was a great physicist, many found him lacking in metaphysics. Here is where Locke enters: "What Newton had not dared to do, or perhaps could not do, Locke undertook and executed with success. One can say that he created metaphysics a bit like Newton had created physics."[28] With the separation of science from philosophy, we now have, by the middle of the eighteenth century, something that looks rather more like our modern intellectual landscape, a domain of science, the study of nature, set off from the domain of the philosopher.

There were, to be sure, important and innovative developments, perhaps even revolutionary things going on in various domains of the study of nature in seventeenth century. But perhaps as important as any of them was way in which what might be called the disciplinary geography changes. An important part of the so-called scientific revolution involves the way the domains in which what we now call science shift and change, the way in which some combine and others build alliances with one another, so that by sometime in the eighteenth century, one can say that

[28] Diderot et al., *Encyclopédie*, vol. 1, p. xxvij.

they form a single confederation, as it were. This involves, among other things, the integration of natural philosophy or physics and mathematics, and the joining of the historical enterprise (the observational study of nature) with the philosopher's attempt to find causes. But there are social, political, and institutional dimensions to the change as well: the foundation of academies and the rise of the institution of the learned journal in which the practitioners of these different approaches to the study of nature can come together and work literally elbow to elbow. This reorientation of the domains of inquiry was a crucial part of the changes that happened in the seventeenth century and resulted in the creation of science as an enterprise. Without it would be impossible to talk about *the* Scientific Revolution in later years.[29]

[29]An earlier version of parts of this essay appeared in Italian as the introduction to *Storia della Scienza, volume V, sezione I: L'età della rivoluzione scientifica*. (Rome: Istituto della Enciclopedia Italiana, 2002).

# The Unity of Natural Philosophy and the End of *Scientia*

**Stephen Gaukroger**

The dominant model of the unity of natural philosophy in the Middle Ages, namely *scientia*, collapsed in the early modern era, and the unity of natural philosophy was rethought in the seventeenth century in terms of a reductionist and foundationalist notion of common causation. I want to argue that, if we can identify the reasons for the collapse of the notion of *scientia*, then we can get a better sense of what was demanded of its successors, and that this will help us in understanding the subordination of all cognitive values to scientific ones so distinctive of the modern era.

## *Scientia* and Its Critics

The Aristotelian procedure in natural philosophy was to work from observed events and processes to underlying principles, and then, starting from these underlying principles, to provide an account of the various forms of physical behaviour that one wanted to explain. It was universally acknowledged that not everything one derived from one's underlying principles was going to be a realised possibility. An account of how things actually behave is not exhaustive of how they might have behaved under any physically-characterisable circumstances, and this is a good part of the rationale for our seeking underlying principles in the first place. In Aristotelian terms, the aim is to provide an account not just of how things actually behave, but of how they would behave in a range of circumstances, thereby transcending local contingencies and focussing on those features of the behaviour of bodies that derive from their essential structure. The second part of the process, the derivation of physical behaviour from these underlying principles, is what the medieval

S. Gaukroger (✉)
University of Sydney and University of Aberdeen
e-mail: stephen.gaukroger@usyd.edu.au

T. Sorell et al. (eds.), *Scientia in Early Modern Philosophy*, Studies in History and Philosophy of Science 24, DOI 10.1007/978-90-481-3077-1_2,

philosophers termed *scientia*,[1] and knowledge consisted in this, not in the discovery of principles, which was a prelude to or precondition for knowledge.[2] *Scientia* was a systematic and encyclopedic form of presentation of knowledge in which known facts were grasped in terms of their underlying principles and causes. Finding out information about the world was a first step, but in itself it did not constitute a form of understanding: knowledge that something was the case was merely a prelude to knowledge why it was the case.

*Scientia* had a number of features that sixteenth and seventeenth century natural philosophers rejected, and as a result the whole idea of *scientia* disappeared, to be replaced by a number of different conceptions of what the domain of natural philosophy consisted in and what held it together. I want to focus on two of the features rejected, both of which hinge on the fact that the unity is achieved by excluding disciplines that might otherwise be thought to fall within the domain of natural-philosophical understanding. In *Metaphysics* E, Aristotle had divided the theoretical sciences into those that deal with whatever is unchanging and has independent existence (God), those that deal with whatever is unchanging but lack independent existence (mathematics), and those that deal with whatever is changing and has independent existence (natural philosophy). The theoretical sciences are distinguished from the practical and productive sciences in that they are concerned not with enabling us to make things or to live well, but with understanding how things are and why they are as they are. This kind of understanding consists in grasping natural processes in terms of the essential principles underlying them, and these essential principles are to be found in the things themselves, for it is from the "natures" of the things themselves that their behaviour derives. It is important to note in this respect that the Aristotelian idea of natural philosophy is tailor-made for essential phenomena, and in consequence of its being tailor-made, it does not fit things for which it wasn't designed. Sixteenth- and seventeenth-century critics focused on two areas where this tailor-made quality caused problems: the exclusion of particular domains from the theoretical sciences, and the exclusion of phenomena on the grounds that they are non-natural processes.

## Contemplative Versus Practical Natural Philosophy

Much sixteenth-century natural history was compiled as a response to perceived gaps in the traditional Aristotelian and didactic classifications of knowledge. The New World contained flora, fauna, and meteorological phenomena that found no

---

[1] Aquinas actually defines *scientia* as the state of mind one is in when one has successfully engaged in this process—see, for example, Aquinas, *Summa theologiae*, Part II, second part, qu. 49 art. 1 and qu. 50 art. 3—but it seems, by extension, to cover the process itself.

[2] See Charles H. Lohr, "Metaphysics and Natural Philosophy as Sciences: the Catholic and Protestant Views in the Sixteenth and Seventeenth Centuries", in Constance Blackwell and Sachiko Kusukawa, eds., *Philosophy in the Sixteenth and Seventeenth Centuries* (Aldershot: Ashgate, 1999), 280–95.

place in the classifications of the ancients, and sixteenth-century Iberians were among the first to question their authority in natural history.[3] At the same time, Northern European thinkers were raising questions about what should and should not be included. Cardano's *De subtilitate* (1550) and *De varietate* (1557), for example, cover natural philosophy and various secrets of the trades and medicine, as well as acrostics, poems hidden in poems, and mathematical conundrums. Johann Jakob Wecker's *De secretis* (1582) moves from the metaphysical and natural-philosophical implications of creation to how to counterfeit coins and gems and how to catch fish. Della Porta's *Magia Naturalis* (1589) deals with many categories usually excluded from classifications of knowledge either because they were considered too ephemeral (the art of beautifying women) or because they cover "marvels" (optical tricks, invisible writing etc.), but he also dealt with practical questions in metallurgy and optics which, if they had been covered in other classifications, were covered inadequately. What lay behind this extension of what was to be included in knowledge was not just the opportunity to challenge the exclusiveness of traditional classifications, but also the opportunity to make clear how far modern knowledge had passed beyond Aristotle. In Book 17 of *De subtilitate*, for example, Cardano notes that there are so many inventions unknown to the ancients—such as domestic furnaces, church bells, stirrups on saddles, counterweights in clocks—that it would take more than a book to list them fully. Sixteenth-century writers in this tradition rejected what they saw as the restrictive classifications employed by Aristotelians, and set out to include many practical disciplines within the ambit of knowledge of the natural realm.

At the beginning of the seventeenth century, Bacon, in his attempt to move natural philosophy from a speculative to a productive discipline, had launched a full-scale assault not only on the traditional classifications, but also on their underlying rationale, which was intimately tied up with their nature as speculative disciplines. In his *Cogitationes de natura rerum*, he criticises Aristotle's distinction between natural motions (rectilinear motions in the case of terrestrial bodies, circular motions in the case of celestial ones) and violent motions. Aristotle had distinguished between natural objects and processes on the one hand, and artifacts and unnatural or constrained or violent processes on the other. Natural philosophy was concerned to explain the properties of things in terms of their essences. What lies at the basis of this conception is the distinction between those things that have an intrinsic principle of change, and those things that have an extrinsic principle of change. An acorn, which has within itself the power to change its state, namely into an oak tree, and a

[3] The question is dealt with in detail in José Antonio Maravall, *Antiguos y Modernos: Visión de la historia e dia de progresso hasta el Renacimiento* (Madrid: Sociedad de Estudios y Publicaciones, 1966). See also Víctor Navarro Brotóns and Enrique Rodríguez Galdeano, *Matemáticas, Cosmología y Humanismo en la España del Siglo XVI. Los Comentarios al Segundo Libro de la Historia Natural de Plinio de Jerónimo Muñoz* (Valencia: Instituto de Estudios Documentales e Históricos sobre la Ciencia Universitat de València, 1998); and more generally Jorge Canizares-Esguerra, "Iberian Science in the Renaissance: Ignored How Much Longer?", *Perspectives in Science* vol. 12 (2004), 86–124: 96–8.

stone raised above the ground, which has the power to change its position, namely to fall to the ground, both come in the first category. In neither case is anything external required for the change/motion to occur. For Aristotle, we explain and understand things by understanding their natures, and to grasp the nature of something is to grasp the source of all its natural properties. If we ask why a stone falls, the answer is that stones are heavy and heavy things fall: that is all there is to it. If we are asked why this tree puts out broad flat leaves in spring and keeps them through the summer, we may reply that it does this because it is a beech. In other words, it is not necessary to look outside the thing to account for its behaviour. And wherever the behaviour of something can be explained without looking outside the thing, that behaviour, and the feature that the thing acquires or retains, is natural. It is natural for stones to fall, it is the nature of beeches to have broad flat leaves. Such explanations are explanations of unconstrained, internally-generated natural processes, and explanations of this kind lie at the core of Aristotle's natural philosophy.[4] Natural philosophy is a *scientia* of natural processes: it tells us why and how they occur. By contrast, there can be no *scientia* of unnatural or constrained or "violent" states and processes, which might be caused by any number of extrinsic events. Natural philosophy cannot be expected to account for these: a stone falling to the ground when released from constraints has a single explanation which refers us to an intrinsic cause, whereas the causes of a stone rising from the ground do not have anything to do with anything essential to the stone, but rather arise from some purely contingent circumstances.

Bacon argues that this way of approaching natural philosophy is fundamentally mistaken. It is "violent" motions, he argues, not natural motions, that should be the subject of natural-philosophical enquiry. These include those unnatural processes produced by mechanical devices such as levers, pulleys, and screws; those strategic unnatural placements of stones that hold buildings up; those unnatural motions of bodies produced by artillery, and so on. These are "the life and soul of artillery, engines, and the whole enterprise of mechanics."[5] Note that Bacon is not concerned with the truth or otherwise of Aristotle's account here. His argument is that the whole investigation is beside the point, so questions of whether it is true are not are simply irrelevant.

Crucial to this approach is his distinction between practical, active enquiry and philosophical contemplation.[6] At the most basic level, what is wrong with

---

[4] There is a good discussion of these issues in William Charlton's introduction and notes to his translation of the *Physics*: *Aristotle's Physics I, II* (Oxford: Oxford University Press, 1970), from which I have taken the example here. For a more comprehensive treatment, see Wolfgang Wieland, *Die aristotelische Physik* (Göttingen: Vandenhoeck & Ruprecht, 1970), and Helen S. Lang, *The Order of Nature in Aristotle's Physics* (Cambridge: Cambridge University Press, 1998).

[5] *The Works of Francis Bacon*, ed. James Spedding, Robert Leslie Ellis, and Douglas Denon Heath (14 vols, London: Longman & Co, 1857–1874), vol. 3, 29 [text]/vol. 5, 433 [trans].

[6] See Stephen Gaukroger, *Francis Bacon and the Transformation of Early Modern Philosophy* (Cambridge: Cambridge University Press, 2001), 44–57.

Aristotle's approach to natural philosophy for Bacon is that it is directed towards seeking a contemplative understanding of natural phenomena.[7] But if one is guided by a concern to pursue natural philosophy with a view to transforming nature for our benefit, it will be wholly inadequate, because it will not deal, or will deal only peripherally, with those natural-philosophical questions than give natural philosophy its legitimacy as a worthwhile area of enquiry in the first place. Bacon distinguishes between understanding how things are made up and what they consist of, on the one hand, and by what force and in what manner they come together, and how they are transformed, on the other. It is the latter that we must seek to understand, he argues, for this is what leads to the augmentation and amplification of human powers. To restrict ourselves to the former is to approach nature as if we were examining "the anatomy of a corpse". We must not concern ourselves with the classification of motions as being natural or violent, as Aristotle did, but investigate instead those "appetites and inclinations and things by which all of the many effects and mutations that are evident in the works of nature and art are made up and brought about."[8] In this way, we will discover and distinguish the different kinds of motion, and then we will be able to hasten or arrest these, and by doing this change and transform matter.

In short, knowledge of why something was the case was restricted to those phenomena that followed from the nature or essence of something, and it excluded any properties or qualities that did not flow from the nature of the thing. This left many kinds of phenomena outside the realm of systematic understanding in the Aristotelian sense, and it was this that precipitated much of the critical response to the Aristotelian approach.

---

[7] The questions here are not as straightforward as they may seem. In the opening paragraphs of the *Metaphysics* ($981^b17$–20), Aristotle seems to take uselessness as a sign of worth: "But as more arts were invented, and some were directed to the necessities of life, others to recreation, the inventors of the latter were naturally always regarded as wiser than the inventors of the former, because their branches of knowledge did not aim at utility". Nevertheless, as Lloyd notes, "The picture of Greek intellectuals as being profoundly unconcerned with the practical applications of their ideas stems largely from texts that belong to a particular, Platonist, tradition." G. E. R. Lloyd, *The Ambitions of Curiosity* (Cambridge: Cambridge University Press, 2002), 70. Note also the dissenting position taken by John Wilkins, one of the main moving forces behind the establishment of the Royal Society, who sees Aristotle as defending practical against speculative knowledge: "And whereas the Mathematicians of those former ages, did possess all their learning, as covetous men doe their wealth, only in thought and notion; the judicious *Aristotle*, like a wise Steward, did lay it out to particular use and improvement, rightly preferring the reality and substance of publicke benefit, before the shadows of some retired speculation, or vulgar opinion." *Mathematicall Magick* (London: Printed by M.F. for Sa. Gellibrand, 1648), 7.

[8] Bacon, *Works*, vol. 3, 20 [text]/vol. 5, 425 [trans]. Bacon is not alone in this approach. There are similar sentiments expressed, for example, in the preface to Guido Ubaldo del Monte, *Mechanicorum Liber* (Pesaro: Apud H. Concordiam, 1577), translated in Stillman Drake and I. E. Drabkin, ed. and trans., *Mechanics in Sixteenth-Century Italy* (Madison: University of Wisconsin Press, 1969), 241–7.

## The Unity of Natural Philosophy

Not only did it leave important things out, however, it failed to account for those things that it did include, in a way that bears centrally on the viability of *scientia*. One point on which early-modern natural philosophers were agreed was the complete lack of explanatory power in purported explanations offered within Aristotelian natural philosophy. "Aristotelians", Hobbes tells us in the first chapter of *Leviathan*, "say, For the cause of *Vision*, that the thing seen, sendeth forth on every side a *visible species* (in English) *a visible shew*, *apparition*, or *aspect*, or *a being seen*; the receiving whereof into the Eye, is *Seeing*."[9] These same Aristotelians, Descartes points out, admit that what they call motion is not well understood: trying to clarify it, they define it as "the act of being which is in potency, in so far as it is in potency", which Descartes finds completely unintelligible.[10] Gassendi agrees. "Great God!" he writes, "Is there a stomach in the world strong enough to digest that? What we were after was the explanation for something very familiar, but this is so complicated that nothing is clear now."[11] And Glanvill similarly has some fun at the expense of Aristotelians:

> Even the most common productions are here resolv'd into *Celestial influences*, *Elemental combinations*, *active* and *passive* principles, and such *generalities*; while the particular manner of them is as hidden as *sympathies*. And if we follow manifest qualities beyond the empty signification of their names; we shall find them as *occult*, as those which are professedly *so*. That heavy Bodies descend by *gravity*, is no better an account then we might expect from a *Rustick*: and again, that *Gravity* is a quality whereby an heavy body descends, is an impertinent *Circle*, and teacheth nothing.[12]

In general, advocates of mechanism had no difficulty showing the circularity and pathological awkwardness of such explanations. Nevertheless, abandoning the idea of invoking essential properties to explain physical behaviour brought new explanatory problems.

There are two issues that I want to draw attention to in this respect. The first is that, in contrast to the early-modern mechanist natural philosophies, Aristotelian natural philosophy does not purport to give an account of all physical phenomena, but only a selection of them. The second is that, again in contrast to the early-modern mechanist natural philosophies, causation is localised in Aristotelian natural philosophy, so that even those phenomena that fall within the domain of explanation are not necessarily causally connected.

---

[9] Thomas Hobbes, *Leviathan or The Matter, Forme and Power of a Commonwealth Ecclesiasticall and Civil* (London: Andrew Crooke, 1651), 4.

[10] Descartes, *Oeuvres de Descartes*, ed. Charles Adam and Paul Tannery (2nd ed., 11 vols, Paris: Vrin, 1974–1986), vol. 11, 39 (*Le Monde*, ch. 7).

[11] Gassendi, *Opera Omnia* (6 vols, Lyon: Lavtentii Anisson & Ioan. Bapt. Devenet, 1658), vol. 3, 186 col. 2 (*Adversus Aristoteleos*, Lib II, ex. 5, art. 4).

[12] Glanvill, *Scepsis Scientifica: or, Confest Ignorance, the way to Science; in an Essay of The Vanity of Dogmatizing, and Confident Opinion* (London: printed by E. Cotes for Henry Eversden, 1665), 126.

Aristotelian natural philosophy had been premissed on the idea that understanding the nature of something enables one to understand its properties. More precisely, its claim was that it enabled one to understand those of its properties that followed from its nature, and it enabled one to understand the behaviour that derived from those properties. The behaviour of bodies that is the result of contingencies in the world, that is, that behaviour that does not derive from something essential to the body, falls outside the domain of natural philosophy. Consequently, there are many physical processes in the world that are not explicable in terms of Aristotelian natural philosophy, and it is not just that these are not explicable in fact, they are not explicable in principle. There is a natural-philosophical explanation for why water pours to the ground when released from constraints, for example, but not for why and how it is lifted by a mechanical device. Aristotelian natural philosophy deals with the natural behaviour of bodies, not any other kind of behaviour they might undergo.

Bacon's primary criticism of Aristotelian natural philosophy, as we have seen, was that it was directed towards understanding of the wrong kinds of processes. What we should be concerning ourselves with are "artificial" processes, those by which we might constrain and control natural phenomena. To meet this demand, what is needed is not a system that excludes natural processes, since these still need to be understood if they are to be controlled, but rather a system that includes both natural and unnatural processes. There are two models for this in the natural philosophies of the Hellenistic era. Stoicism offered a holistic view of the cosmos in which everything is in effect natural, for everything is connected with everything else: the universe of beings is conceived as an immense physical organism organised around an internal principle that, by the analogy with living beings, is vital spirit or *pneuma*. This kind of model, which became popular in eclectic Neoplatonist and Aristotelian versions in the sixteenth century, was largely abandoned after then. Epicureanism, by contrast, offered a picture with no such intrinsic connections but in which the ultimate constituents of the universe, atoms, exhibited an absolute uniformity of behaviour (leaving to one side the problematic "swerve", which was abandoned by early-modern advocates of atomism). On the Epicurean model, extrinsic connections fulfill the same function that intrinsic ones do in the Stoic picture, so that in both cases there is a single fundamental level at which everything physical could be accounted for, and the cosmic system is deterministic.

This is in contrast to earlier systems, those of Plato and Aristotle. For Plato, exceptionless regularity is to be found in the Forms rather than in sensible phenomena. The classic notion of determinism—of a system in which every state of affairs is a necessary consequence of a chain of preceding causes—is almost entirely absent from Aristotle's approach also. More important for Aristotle is the contrast between the absolute necessity and invariance that applies to the motions of heavenly bodies, to mathematical truths, and to certain attributes of being in the sublunary world such as human mortality and, on the other hand, the irregularity and variation of many aspects of the sublunary world, where the most that can be said of many things is that they happen for the most part but not always, and where there are many accidental connections that fall outside the scope of scientific knowledge altogether, where scientific knowledge is concerned with what is always or usually the case. Aristotle's picture of the consequences of an event is not one of chains of cause and effect interwoven in a nexus extending to infinity, but rather, in an analogy used by David Balme, one resembling the ripples caused by the throwing of a stone into a pond, which spread out and combine with the ripples caused by other stones, but eventually die away and come to nothing.[13] And conversely, Aristotle can assert that there are fresh beginnings (*archai*), not confined to human agency, without supposing that there is a deterministic causal nexus occasionally interrupted by undetermined events, for he simply does not see the question in these terms.[14]

The relation between explanation, causation, and necessitation is complex on the Aristotelian approach. On the Stoic and Epicurean accounts, by contrast, it is straightforward. In the latter, we assume that, at some level of description, everything in the universe is causally connected with everything else: as a consequence, there are no events for which they are no connecting causal chains, provided we are prepared to search back far enough through causal ancestors. Explanations are simply statements of cause, and causes always necessitate their effects. Indeed they are treated as if they were simply sufficient conditions for the effect. The Stoics postulated intrinsic connections between events on the model of the relation between the organs of a body, whereas Epicurean atoms are quite independent of each other in the sense that, except when they are involved in collisions, their behaviour is unaffected by any other atom. For Aristotle and the Stoics, bodies act differently depending on their natures, whereas for the Epicureans, bodies have no individual natures. They are completely homogeneous, as we would expect in the case of things having no individual natures, and that is why their fundamental behaviour is the same.

## Reduction

What unifies all physical phenomena on this view is the idea that their behaviour is due to the behaviour of the microscopic corpuscules that make them up.[15] Combined with the idea that there is no behaviour of physical things that cannot be

---

[13] David M. Balme, "Greek Science and Mechanism I", *Classical Quarterly* vol. 33 (1939), 129–38.

[14] See the discussions in Richard Sorabji, *Necessity, Cause and Blame: Perspectives on Aristotle's Theory* (London: Duckworth, 1980); and Sarah Waterlow, *Nature, Change and Agency in Aristotle's Physics* (Oxford: Oxford University Press, 1982).

[15] It should be said that, from an observational and empirical point of view, this was always a wholly speculative hypothesis. It fared particularly badly in the wake of the development of microscopy, primarily because it was fundamental to micro-corpuscularianism that complexity was a feature of the macroscopic world, whereas at the microscopic level processes were extremely simple and economical: what the microscope actually revealed, however, as is evident from Hooke's *Micrographia*, was a world not with less complexity but with significantly more complexity. See Catherine Wilson, *The Invisible World* (Princeton: Princeton University Press, 1995).

accounted for in this way, the principle of common cause unifies the realm of natural-philosophical enquiry in a way that does not exclude either "artificial" phenomena such as those studied in mechanics, or any other kind of physical phenomenon, whether or not it falls under Aristotle's idea of the "theoretical sciences".

The corpuscularianism that replaced Aristotelian matter theory substituted, for the Aristotelian notion of cause as something that acts locally, the idea that there is a level of causation common to all natural phenomena, namely the microscopic level of atomic collisions. This acted to unify the natural-philosophical enterprise because, by contrast with the Aristotelian model, there was a fundamental level of causal interaction in which they all participated, and which shaped the behaviour of the whole physical system. But the level of common causation has two problems. In the first place, its explanatory value is very limited, and it is unclear what one should do about experimental explanations that identify well-supported causal processes with no apparent connection to any underlying corpuscularian account. Second, the level of common causation should, ideally, identify a small number of simple primary qualities which are fully quantified, and the interactions between corpuscles having only these qualities should then give rise to the complex macroscopic world of secondary qualities. But in fact the reverse is the case: it is the macroscopic world that yields to quantification, the microscopic processes remaining in the realm of speculation.

Both these issues come to the fore in the disputes between foundational mechanists and advocates of experimental philosophy. In the dispute between Boyle and Hobbes over the air pump, for example, Boyle insisted on drawing connections between phenomena at the phenomenal level, rather than seeking to ground phenomena in underlying causes, whereas Hobbes insisted that the former could have no explanatory value in their own right. Yet appeal to underlying causes could give no account of these phenomena in the case of pneumatics—the production of a transparent space on the evacuation of air from a container made no sense in Hobbes' mediumistic account of the transmission of light. Nor could any appeal to underlying causes fault the well-attested results that Boyle produced. Newton's report of his prism experiment, in which the heterogenity of sunlight is shown, is even more problematic for foundationalist mechanism. In his pioneering work on the production of colour, Descartes had provided a geometrical account of how sunlight, on being refracted at a particular range of angles, produces spectral colours. But once he moves to his micro-corpuscularian account of how these colours arise, he shifts register as it were to an account in which geometrical and quantitative elements are absent in any intrinsic way. The idea is that, on being refracted through a medium at a particular angles, the corpuscles making up the refracted ray are moved at different rates at either side of the ray because of the different angles at which the extremities of the light ray emerge from the refracting interface. This sets up a series of rotations in the corpuscles, which decrease in speed as one moves from one end of the ray to the other, and we respond visually to these difference in speed of rotation by perceiving different colours. The speed of rotation is something entirely speculative and qualitative however: qualitative, in that the differing speeds are just correlates of the colours seen, and have no independent rationale. In other words, the corpuscularian reduction takes us away from a quantifiable realm into a speculative qualitative one. Compare this with Newton's procedure. By confining himself to changes in the shape of the refracted image in a strategic conjunction of prisms and boards, Newton is able to keep it as part of an exercise in geometrical optics, thus ensuring that we do not leave the quantitative realm. The phenomena can thus be connected mathematically, rather than in terms of a matter-theoretical account of underlying physical processes that was a *sine qua non* of traditional natural philosophy.

A second issue for mechanist reduction, is the question of how the microscopic/macroscopic distinction functions when taken out of its original home of matter theory. By the middle of the seventeenth century, the behaviour of fundamental corpuscles was being accounted for primarily in mechanical rather than matter-theoretical terms, and whether processes are microscopic or macroscopic is irrelevant in mechanics. Here we get one a crucial shift in reductionist thinking: the replacement of the microscopic/macroscopic distinction with the primary/secondary one, and with this a way of proceeding that is more in tune with the requirements of mechanics. This provides a means of doing something that had always been at the core of corpuscularianism from Democritean atomism onwards, namely identifying what might be termed a fundamental level of reality. Democritus had advocated that the microscopic world is more real than the macroscopic one, indeed that it was the only true reality. Ultimately, all that existed were atoms and the void. But even his successors in the atomist tradition considered this implausible, and none of them restricted reality to the microscopic level. Epicurus, for example, while insisting that everything is composed of atoms, did not deny reality to combinations of atoms, or to the distinctive emergent sensory qualities that such combinations took on, qualities that were not to be found at the atomic level. But if it could not plausibly be argued that the microscopic world was more real than the macroscopic one, it *could* be argued that the world of primary qualities was more real than that of secondary qualities: that primary qualities were really in the world whereas secondary qualities were merely psychic additions of the perceiving mind. Natural philosophy thus becomes focused on the former, and since the realm of primary qualities is what natural philosophy deals with, it is to this extent unified in a substantial sense.

## What Cost Unification?

The foundationalist mechanism that I have described—and which I associate in its seventeenth-century development with figures such as Descartes, Hobbes, Huygens, and Malebranche—is committed to the idea that mechanics contains the only explanatory resources available to natural philosophy, which effectively requires a reduction of natural philosophy to mechanics. This view was encouraged by the explanatory success of Newton's *Principia*, where many of the kinds of results that mechanists had tried to achieve were supplied by an approach which dealt with the problems in an unexpected way, failing to resolve some issues that most considered had to be resolved before the question could be pursued seriously, yet producing a quantitative unification of the fields of terrestrial and celestial mechanics. The increase in the explanatory power of mechanics that the *Principia* represented was a significant factor in the establishment of mechanism as a world view. With this came a completely different understanding of the unity of natural philosophy, and more generally the unity of knowledge, than that fostered in the *scientia* conception.

Yet substituting reductionism for essentialism was not the only possible response to the collapse of *scientia*. As well as reduction, there is the strategy of accommodation. We can think of the two options in terms of two metaphors: that of a tree and that of a box of tools. The reductionist will aim at a single all-inclusive account of natural phenomena, arranging these phenomena systematically in what can essentially be represented in a tree-like form which reflects supposed ontological and explanatory priority. The foundations on which the system rest, in seventeenth-century natural philosophies, tend to be reductionist rather than essentialist, but the key part of the exercise is tracing back the physical behaviour of bodies and their interactions to first principles, because this is where the explanatory power lies. Of course there will often be disagreement over the nature of these first principles, but there is agreement that natural phenomena form a single closed system and that the ultimate aim of natural philosophy is the discovery of the single underlying structure on which all these phenomena rest, and which provides them with a coherent structure.

What drives accommodation is very different from this. Consider the metaphor of the tool box. The tools are all designed to work a particular kind of material—say wood—but they are relatively refined. Each of the tools can be used for some things and not others, even though there may be some overlap in uses, and tools may evolve, with earlier models being replaced by new ones, which may have more uses than the one they replace (although they may equally have a more specialised use). If one faces a particularly intractable problem one may devise a new tool only distantly related to anything already in the box. Of course it is important that one keep the contents of the toolbox to a manageable size, because a proliferation of tools may be unproductive. One might prefer a single tool that has a wide range but performs some operations better than others to many very specialised tools that give marginally quicker results, on the grounds of economy. But there is no rationale for thinking that one might reduce them all to one basic tool or one species of tool, nor is there any basis for ranking the tools in a hierarchy from more to less fundamental. And indeed in the case of especially intractable problems, one is not prevented from using anything that comes to hand to do the job. Boyle and Hooke are paradigm examples of this kind of approach in natural philosophy, but so too is Newton.

As I have indicated, neither Boyle in his pneumatics nor Newton in his account of colour believed themselves constrained by what kind of rationale could be given for their results by a foundational corpuscularianism. Quite the contrary, the requirement was in the other direction: it wasn't that Boylean pneumatics had to accommodate itself to foundational corpuscularianism, but rather that foundational corpuscularianism had to account for Boyle's results. The results were not only treated as secure in their own right, but as being independent of whatever corpuscularian rationale might be provided for them. Their explanatory power is diminished radically once one tries to treat them in corpuscularianism terms, and indeed this is a manifestly artificial move, for this explanatory power does not depend on how one characterises the material constitution of gases or of light.

There are some very basic issues at stake here, and two questions in particular come to the fore: Would it matter if some over-arching structure could not be imposed on natural philosophy? Would it be a problem if the attempt to provide such an overarching structure prevented genuine developments at the phenomenal level?

The first question envisages a piecemeal understanding of natural phenomena and the problem is whether a form of non-systematic understanding is genuine understanding. This issue divides advocates of experimental and foundational natural philosophy. One response open to the former is that it is a matter of a number of competing natural philosophies which act to regulate natural-philosophical discussion, so that while it is true that one has to choose some system if one is to have some kind of anchorage in these disputes, this is an anchorage that nevertheless can be shifted as circumstances dictate. In other words, it is the experimentally-established cases that carry the explanatory weight and hence it is these that do the real natural-philosophical work, not the systems in which one decides to anchor them, which will have to be revised or jettisoned if they fail to cohere with these. I see not only Boyle, Hooke and Newton in this tradition but also many of the eighteenth- and nineteenth-century encyclopedists from Diderot onwards.[16]

By contrast, the best case to be made for the foundational path is effectively that made by Descartes and, following him, Huygens. The argument is that attention has to be paid to whether purported explanations actually explain, a crucial issue in the rejection of Aristotelian natural philosophy, where generations of natural philosophers had been committed to accounting for natural phenomena in a way which, in the view of the advocates of mechanism, could not possibly offer enlightenment. What is needed, therefore, is a way of securing that purported explanations actually do explain. The Cartesian claim is that, without the characteristics of clarity and distinctness, any purported explanation is impossible, because we cannot identify exactly what it is that is being proposed in the explanation. In other words, the lesson Cartesians draw from the past is that any reform of natural philosophy must begin with some precise criteria for what is going to count as a satisfactory explanation, and, whatever extra criteria might be introduced to deal with specific areas, these must hold generally in any kind of cognitive endeavour. This results in a unification of cognitive enquiry, and it turns out it is one with a mathematical model to the extent that mathematics embodies the qualities of clarity and distinctness in an archetypical way. So as well as unifying cognitive enquiry about the natural world, it also introduces a cognitive hierarchy by which those disciplines that can be

---

[16] This raises the question of the difference between Renaissance and Enlightenment encyclopedism, one of the most striking features of which is the shift from presenting material starting from first principles to alphabetical ordering, reflecting a wholly different conception of the point of the exercise.

quantified exhibit a degree of clarity and distinctness, and hence explanatory power, which is greater than that of other areas.

Notice two things about this approach. First, it is a response to a specific set of problems: the vacuous nature of explanation in Aristotelian natural philosophy, the question of the degree of certainty to be achieved in physical enquiry, the ability to make discoveries in a systematic way rather than simply coming across them by accident, and so on. If the problems identified in the Aristotelian approach are genuine ones—and while the situation is complicated by factors such as differences in what requires explanation in Aristotelian and mechanist natural philosophies, the claim is certainly not baseless—then the failing is not only serious but fatal, something that it would be wise to resolve before proceeding to construct an alternative natural philosophy. Moreover, there was widespread agreement on this diagnosis of the weakness of Aristotelian natural philosophy: there can be no doubt that it is a view that Boyle, Hooke and Newton shared with Descartes and Huygens, for example. The Cartesian approach certainly deals with the question directly, but this in itself doesn't guarantee that it is the right solution. The notions of clarity and distinctness work well in the context of mathematics, but where our assessment of a theory depends on empirical questions their centrality is not quite so evident. And it is on such empirical questions that issues concerning the air pump and Newton's prism experiment were disputed. Moreover, within the confines of the experimental data that they produce, Boyle and Newton cannot be faulted on their explanations, which are the only ones of the available options that succeed in accounting fully for the phenomena. The charge that they are not satisfactory when taken in a broader context of foundationalist natural philosophy is not plausible in these cases. Quite the contrary, it is the foundationalist natural philosophy that has to adjust to accommodate these results. This doesn't settle the question of course, because a foundationalist might in fact be able to accommodate these results in a satisfactory way. The point is rather that the provision of a foundationalist natural philosophy is not the only way of meeting the requirement that explanations be genuinely explanatory, but that in the cases of the air pump and prisms, the onus is on the foundationalist—as Hobbes learned to his cost—to show that basic corpuscularianism had the requisite explanatory resources.

Second, mathematics is treated as the archetypical case of clarity and distinctness, and this makes things turn on mathematics to a considerable extent. The fact that, for the Cartesian tradition, the criterion of clarity and distinctness is paradigmatically manifested in mathematics, means that those parts of physical theory that can be translated into mathematical terms have a foundational priority, and it is this kind of consideration that motivates Huygens' attempt to pursue natural philosophy purely in terms of kinematics. However, Newton's account of the production of colour not only offers a case of successful quantification, but offers it in a situation where the corpuscularian account offered by Descartes actually takes us away from the quantified phenomena into the qualitative realm of matter theory. In Cartesian terms, this should mean that the purely phenomenal account generated in "experimental" natural philosophy has a claim to being more clear and distinct than the more fundamental corpuscularian one. What Newton's prism experiment suggests is the possibility that the mathematics might best be done at the phenomenal level.

The second question is whether it would it be a problem if the attempt to provide an overarching structure prevented genuine developments at the phenomenal level. The issue is a real one, as the response of Huygens to Newton's work in optics and dynamics shows. It is not just that these could not be a complete accounts for the matter for Huygens, and must be explicated in more fundamental mechanical-corpuscularian terms, but that his kinematically-driven trajectory of research takes him in a direction where, for foundational reasons, he is obliged to pursue a programme of kinematic reduction which had rapidly diminishing returns. On the other hand, Huygens would see natural philosophy as losing crucial guidance if the search for overarching structure of a particular kind were abandoned. Particular successes, even on the scale of the mechanics of the *Principia*, can still be compatible with a programme that is moving in the wrong direction: Ptolemaic and Tychonic astronomy would doubtless have come to mind for seventeenth-century natural philosophers.

## Conclusion

The issue, then, is not as simple as an essentialist *scientia* model being replaced by a reductionist mechanist one, where it is taken as given that the domain of natural-philosophical enquiry requires an underlying principle of unity, and the exercise is to discover what this principle of unity is. We can also question the need for such a principle of unity, and this has ramifications that go beyond the early-modern abandonment of the notion of *scientia*.

The assimilation of all cognitive values to scientific ones is one of the overarching features of modernity, and the single most important ingredient in the consolidation of such a scientific culture in the West is the idea of the "unity of science". Opinions on what this unity consists in have differed, but however it is conceived, without some viable understanding of the "unity of science", the idea that cognitive values generally can be assimilated to scientific ones will collapse. If "science" is simply a loose grouping of disciplines with different subject matters and different methods, tied in various ways each of which work for some purposes but not for others, then there can be no modelling of cognitive values generally on scientific ones. Contrary to the approach of the Logical Positivists, for example, which was characterised by a combination of wishful thinking and a priori arguments about a highly idealised picture of science,[17] there can be no purely conceptual way of deciding this issue, no question whether, independently of whatever strategies might have been at play in particular periods of the development of science, science is an intrinsically unified enterprise. We need to know what unification strategies have been

[17] See, for example, the contributions to vol. 1 of Otto Neurath, Rudolph Carnap, and Charles Morris, eds., *Foundations of the Unity of Science* (2 vols, Chicago: University of Chiaco Press, 1970). Neurath's opening essay, "Unified Science as Encyclopedic Integration", 1–27 (first published 1938) gives the flavour of the project.

employed: whether they are methodological or involve a reduction of disciplines to some foundational discipline. We need to know what motivates them: whether they are designed to counter explanatory failure, for example, or to secure results in a systematic rather than an accidental fashion, or to provide guidance about the general aims of enquiry, or to make manifest some underlying unity that is obscured by the various forms of experimental enquiry that one is constrained to follow. We need to identify just what disciplines are being subjected to this unifying strategy, and what the model of unification or reduction is. Without asking these questions, it is clear that any consideration of "the unity of science" is premature.

# Matter, Mortality, and the Changing Ideal of Science

**Catherine Wilson**

## I Gassendi's Attack on Aristotelian *Scientia*

"*Nihil sciatur*" said Gassendi, echoing the sixteenth century sceptics Michel Montaigne and Francisco Sanchez, and he proceeded to attack, in his unfinished *Exercitationes* of 1624, the edifice of Aristotelian physics, metaphysics, theory of the soul, and teachings on generation and corruption. The notions of matter, form, and privation, he declared, were no basis for the knowledge of essences to which Aristotle thought the sciences aspired. "It is manifestly clear that so far we know nothing about natural things through the efforts of all philosophy." Take the flea, for example, a common enough animal:

> You say that the flea is composed of matter and form, and that the privation of this form preceded the generation of this flea in this matter....Oh what an unnourishing and indigestible philosophy! ....I desired to know just what sort this matter was, what dispositions it required to receive that form, for what reason it was distributed so that this part of it went into the proboscis, that part into its feet, another into its hair and scales...what was the active force and how was it brought to bear when it formed both the entire body and its very different parts....[1]

Two central themes, evident in the quoted passage, dominated seventeenth century epistemology: first, the inanity or falsity of most of what had passed for knowledge of nature and civil society—"philosophy," in contemporary, though not our parlance; and second, the superiority of explanations citing material conditions and efficient causes over explanations citing incorporeal entities or agents.

---

C. Wilson (✉)
School of Divinity, History and Philosophy, University of Aberdeen
e-mail: c.wilson@abdn.ac.uk

[1] "... By what power does the flea bite you so sharply to ingest his nourishment from you, how does he digest it and assimilate part of it in various passages, and transform part of it into spirits which conserve him and impart life to his entire body, and eliminate its superfluous parts through his different winding intestines....?" *Exercitationes Paradoxicae Adversus Aristoteleos*, Bk II, Ex. VI, 6, tr in *The Selected Works of Pierre Gassendi*, tr. and ed. Craig Brush, NY and London, Johnson Reprint, 1972, p.98.

T. Sorell et al. (eds.), *Scientia in Early Modern Philosophy*, Studies in History and Philosophy of Science 24, DOI 10.1007/978-90-481-3077-1_3,

To refashion a conception of scientific knowledge, early modern philosophers reverted to explanatory schema not entirely foreign to Aristotelian natural philosophy, but of whose limited applicability Aristotle was firmly convinced. Adequate explanations, especially in the biological realm in which he specialized, should cite not only material conditions and efficient causes, but final and formal causes. Though he rejected the Platonic doctrine of separated forms and was unclear about the existence of human souls uncoupled from human bodies,[2] Aristotle recognized souls, forms, and divinities as active participants in natural processes. These incorporeal entities moved celestial and animal bodies, shaped matter in generation and lost control of it in corruption, as well as endowing formless, primary matter with its various characteristics. Moreover, though immaterial, these active agents were knowable. Philosophical analysis discovered them and the natural scientist studied their manifest workings. "The study of the soul must fall within the science of nature."[3] Aristotle had memorably insisted on the inanity or falsity of the natural philosophy of his materialistic predecessors, and their concentration on material and efficient causes. "Now, the causes being four, it is the business of the student of nature to know about them all, and if he refers his problems back to all of them, he will assign the 'why' in the way proper to his science—the matter, the form, the mover, that for the sake of which."[4] Moreover, the "genuine physicist" is one who does not confine himself to the material aspect of things. "The [genuine] physicist is he who concerns himself with all the properties active and passive of bodies or materials thus or thus defined."[5] His predecessors had gone astray by mostly ignoring the contribution of incorporeals:

> If we look at the ancients, natural science would seem to be concerned with the matter. (It was only very slightly that Empedocles and Democritus touched on form and essence.) But if on the other hand art imitates nature, and it is the part of the same discipline to know the form and the matter up to a point (e.g. the doctor has a knowledge of health and also of bile and phlegm, in which health is realized and the builder both of the form of the house and of the matter, namely that it is bricks and beams, and so forth)....it would be the part of physics also to know nature in both its senses.[6]

Aristotle's antimaterialistic observations and arguments, comprising mathematical objections to least elements, criticisms of reductionistic theories of qualities, and a dismissal of purely mechanical theories of animation and generation, were, taken together with medieval Christian doctrine, a formidable defense against any science that proposed that ultimately there were only atoms and void, matter and motion.

---

[2]"It indubitably follows that the soul is inseparable from its body, or at any rate that some parts of it are (If it has parts), for the actuality of some of them is nothing but the actualities of their bodily parts.... yet some may be separable because they are not the actualities of any body at all." Aristotle, *On the Soul* 413a, tr. J.A. Smith.

[3]Ibid., 403a.

[4]Aristotle, *Physics* 198a tr. R.P. Hardie and R.K. Gaye.

[5]Aristotle, *On the Soul*, 403b.

[6]Aristotle, *Physics* 194a.

The seventeenth century reversion to the system of Democritus and his successors, Epicurus and Lucretius, accordingly poses some interesting questions.

The recovery and extension of alternative traditions of inquiry that Aristotle had scorned or neglected, or merely left undeveloped was stimulated by and responsible in turn for renewed attention to Epicurus's teachings and Lucretius's Latin poem *De Rerum Natura*. It is reasonably well understood how, with the help of these ancient texts, seventeenth century philosophers refashioned the image of nature in corpuscularian terms, refashioned the notion of explanation in mechanical terms, and began to try to refashion epistemology to accommodate the theories of perception and knowledge that followed from them. The innovators did not only dispatch Aristotle's mathematical arguments against atoms, and ignore most of his pleas for final causes, they rediscovered ancient materialism, ancient mechanics, and the experimental practice of the Alexandrian school in medicine; they improved upon optics, astronomy, mathematics, and physiology. Action at a distance, occult influence, and contagion were now conceived as processes involving subvisible but robustly material corpuscles—at least until Newton raised doubts as to the necessity of such accounts and their application everywhere. To be sure, to those versed in the literature of the ancients, as well as to those who had no taste for academic philosophy, the domination of the curriculum by Aristotle had always appeared somewhat arbitrary: the "innovators" or "moderns" of the seventeenth century followed in the footsteps of well-read humanists in this respect.

Some aspects of the renewal and expansion of the pre-Aristotelian and Hellenistic programmes have been little studied, and this paper will give some attention to the question of the immortality of the human soul in mid to late seventeenth century English natural philosophy. While radicals like Hobbes were prepared to banish all incorporeal entities and spiritual substances, including God, from philosophical discourse, or to re-analyze them in material terms, few philosophers were prepared to go this far. Locke's magnificently evasive strategy with respect to the question of morally significant incorporeals is a key to the interpretation of his *Essay Concerning Human Understanding*. It is one of the central elements of his efforts to lay the foundations of a Gassendist science of appearances, not a critical idealism, but certainly a "critical ideaism."

## II Atomism and the Mortality of the Soul

Gassendi's devaluation of Aristotle depended upon a favourable re-elevation of the Epicurean appeal to experience as the touchstone of knowledge, as well as the Epicurean theory of the atom. The atom—a "material principle," "the primary and universal material of all things" was rightly held in his view to be the ground of all variety in objects—rarity and density, softness and hardness, the ground of sensation in animals, and the cause of generation which, according to Gassendi, depended on molecular seeds.[7] Long calumnified and partially suppressed, Epicureanism flour-

[7] Gassendi, *Syntagma philosophiae Epicuri*, The Hague, 1659; London, 1660, Sect I, Bk III. Ch. 8.

ished in the libertine circles of Paris, on whose fringes Gassendi lived. Members of these cults, according to one of their critics, Pere Garasse, subscribed to the following two theses:

1. There is no other divinity or supreme power in the world except Nature, who requires to be pleased in everything, refusing nothing to our bodies and our senses of what they desire from us in the exercise of their powers and natural faculties.
2. Assuming there is a God, as it is well to maintain to avoid conflict with ordinary superstitious people, it does not follow that there are purely intellectual creatures separated from matter. Everything in nature is a composite. There are neither angels nor devils in the world and it is not assured that the soul of man is immortal.[8]

The last article was especially significant. Lucretius had claimed that "Death ... is nothing to us and does not affect us in the least."[9] The human mind is as material as everything else. It "is exceedingly subtle, being composed of the minutest particles."[10] It is distributed throughout the entire body: "The entire soul is composed of very small seeds, which form a chain through the veins, flesh and sinews. . . ."[11] While the soul atoms are qualitatively different from those composing the rest of the body, the Epicurean teaching is that the soul is not sentient except "in virtue of the neighborhood and interconnection" between itself and the body.[12] It is accordingly impossible to experience without an intact body, or to experience one's own demise. The nocturnal anxiety of the mind brooding over death is dispelled, Lucretius said, "not by the sun's rays and the dazzling darts of day, but by the study of the superficial aspect and underlying principle of nature."[13] Light verse did not recommend so much the study of underlying principles as a philosophy of *carpe diem*.[14]

For non-libertine philosophers, it was a relatively simple matter to situate the corpuscularian image of nature and the new requirements for explanation and criteria for knowledge within a framework of creationism, and to posit God as the author of the laws of nature in the new mathematical sense they were acquiring. Empedoclean materialism had been condemned by Aristotle as incapable of accounting for natural regularities, and Democritean atomism, with its blind, falling, and randomly

[8]Antoine Adam, ed., *Les libertins au XVIIe siècle*, Paris, Buchet/Chastel, 1964, p. 42.

[9]Lucretius, *On the Nature of Things*, III: 830. tr. and ed. Martin Ferguson Smith, Indianapolis, Hackett, 2001.

[10]Ibid., III: 179–180.

[11]Ibid., III: 216 f.

[12]Epicurus, in Diogenes Laertius, *Lives of the Eminent Philosophers*, 2 vols., tr. R.D. Hicks, Cambridge, MA, Harvard University Press, 1972, X: 63–65.

[13]Lucretius, *On the Nature of Things*, II:60 ff.

[14]"A quoy bon tant craindre/Les horreurs du tombeau/Quand on voit eteindre/De nos jours le flambeau? L'ame est une etincelle/Et tout ce qu'on dit de l'esprit/Est bagatelle," Charles Blot, reproduced in Adam, ed., *Libertins*, p. 84.

colliding atoms seemed inadequate to the task of forming and sustaining the visible world. Theomechanism provided a ready solution to the difficulty. Gassendi pointed out that corpuscularian mechanism implied a divine regulator of the atoms, and he claimed that atomism was better adapted to theism than Aristotelianism had been.

The Epicurean teaching on the mortality of the soul presented, however, a more difficult problem than the existence of God—agreed to by everyone, except "the fool"—the libertine victim of motivated irrationality—and his creation and supervision of nature. The claim that death is final, irrevocable and the end of all experience had been the focus of Christian opprobrium for centuries. If it was true, Christianity, with its story of Christ's sacrifice and the promise of eternal life was a fairy tale, and the threat of divine vengeance was empty. The existence of an immortal human soul was, however, very difficult to establish. Neither introspection nor a survey of the magnificence of the visible world tended to its verification.[15] It was a tenet of revealed religion, a *promise* of the New Testament, not even to be found in the Old. The only ancient philosopher who had advanced a credible argument for the immortality of the soul was Plato; Aristotle, as noted, seemed to waver, as was consistent with his hylomorphic commitments. Given a commitment to *corpus et inane*, or at least to *corpus* without forms, as constitutive of nature, and to reason and experience as the touchstones of truth, how was it to be asserted? Perhaps "God" and "the human soul" were fictions generated by material brains and sustained by custom, tradition, and authority—in a word, superstition—as the Epicureans maintained. Descartes announced that he would actually demonstrate the immortality of the soul as well as the existence of God in his *Meditations*. His "method of doubt" elicited the conclusion that the soul was incorporeal and did not depend on the body, and he insisted that there was no reason to suppose that an incorporeal soul that could survive the death of its body would not do so. He tried to show that God, though most certainly this idea in the mind, as materialists maintained, was not *merely* this idea in the mind, but a special kind of idea whose very existence implied the non-fictivity of its object.[16]

This alleged demonstration was greeted with skepticism even on the part of good Christians like Arnauld and Mersenne. Anyone could see that the *immortality* of the soul did not actually follow from Descartes's claim that the soul was a substance distinct from the body. Offering weak demonstrations was no answer to libertines, and presumably worse than not even trying to answer them. The Sixth Set of Objectors—miscellaneous philosophers and theologians—cited St. Paul and the author of Ecclesiastes as skeptics, insisting that Descartes had not met an acceptable standard of demonstration, and that, he had not shown that "corporeal motions" were

---

[15]Indeed, Lucretius offers many observations in favour of the materiality and perishability of the soul; the incompetence associated with youth and old age, *Nature of Things* III:455 ff., and with drunkenness, III:476; the effects of injury to the pupil of the eye, III:408 ff., and blows to the sense organs; and the fixed instincts of various species of animal, III:741 ff.

[16]Rene Descartes, *Meditations*, in *Oeuvres*, 11 vols., ed C. Adam and P. Tannery, J. Vrin, 1965, VII: 46 ff.

inadequate to produce thought.[17] He had not shown that humans possessed faculties above and beyond those of beasts, in whom a corporeal and mortal soul is posited. "Since we do not know what can be done by bodies and their motions, and since you confess that without a divine revelation no one can know everything which God has imparted or could impart to any object, how can you possibly have known that God has not implanted in certain bodies a power or property enabling them to doubt, think etc.?"[18]

Gassendi himself was circumspect and ambiguous on the topic of the human soul. Though he posited and defended a human soul that was non-atomic, incorporeal and immortal, in addition to a material animal soul, composed of very fine, subtle, fiery atoms, the "flower of matter," there is doubt about Gassendi's actual loyalties, especially as he urged an Epicurean "effluvial" soul, composed of dispersible atoms, against Descartes's thinking substance and denied that there was any important distinction between animal and human cogitation.[19] Moreover, he said,

> You can exist apart from your solid body just as vapour with smell can exist outside solid apple. Indeed supposing you are some corporeal or tenuous substance, you would not be said to vanish wholly at your death or to pass into nothingness, you would be said to subsist by means of your dispersed parts. We would however have to say that because of this dispersal you would not continue to think, or be a thinking thing, a mind or a soul.[20]

## III Mortalism: Epistemological Considerations

While Thomas Hobbes, Margaret Cavendish, and Walter Charleton[21] have been variously credited with the importation of Gassendi's materialism into mid-century English philosophy, R.H. Kargon has pointed to a pre-existing tradition of Epicure-

---

[17] "No one has yet been able to grasp that demonstration of yours by which you think you have proved that what you call thought cannot be a kind of corporeal motion. Have you used your method of analysis to separate off all the motions of that rarefied matter of yours? Is this what makes you so certain? And can you therefore show us....that it is self-contradictory that our thoughts should be reducible to these corporeal motions?" AT VII:413; tr. in *Philosophical Writings*, 3 vols. tr. and ed. J. Cottingham, R. Stoothoff, D. Murdoch, and A Kenny, Cambridge: Cambridge University Press, (1985–1993), II: 278.

[18] Ibid., AT VII:421, *Philosophical Writings*. II: 284.

[19] On Gassendi's theory of the soul, see Emily Michael, "Renaissance Theories of Body, Soul, and Mind," in *Psyche and Soma: Physicians and Metaphysicians on the Mind-Body problem from Antiquity to Enlightenment*, ed. P. Potter and J. Wright, Oxford University Press, 2000, pp. 147–173; "Gassendi's Method Illustrated by His Account of the Soul," in *Pierre Gassendi 1592–1992*, 2 vols., Digne-Les-Bains, Société Sciéntifique et Littéraire des Alpes de Haute-Provence 1994, I: 181–193.

[20] Gassendi, *Fifth Set of Objections*, AT VII: 343, *Philosophical Writings*, II: 238.

[21] Some important events in the rehabilitation of Epicurus in England include the publication of Margaret Cavendish's *Poems and Fancies* in 1653; Thomas Stanley's translation of parts of Gassendi's *Syntagma* in his *History of Philosophy* in 1655, John Evelyn's publication of his translation of the first book of *De Rerum Natura* in 1656, and Boyle's *Origin of Forms and Qualities* in 1664.

anism that he locates in Sir Walter Raleigh's band of free-thinkers and in the circle around Henry Percy,[22] and Christopher Hill has emphasized the importance of long standing verbal traditions of heterodoxy, including materialism and mortalism.[23] In the upheavals of the Civil War, many unusual books and pamphlets found their way to the press or into the country from abroad. In 1644, the English Leveller, Richard Overton "the pamphleteer of Amsterdam" published a tract, *Mans Mortalitie*, that was expanded and reprinted several times. Overton offered a battery of arguments and scriptural citations against the "Fancie of the Soul." Apart from the citation of many ancient authorities, including Pliny, on the mortality of man in all respects, the general line of argument was this: the soul is nothing over and above the faculties of man, including reason, consideration, and science. These faculties differ in their degree of perfection but not in kind from those of beasts, as a result of learning and education. They are "temperatures" in beasts, depending on "corpulent matter," so they must be so as well in man. If they are subject to extinction, then "the invention of the Soule upon that ground [the immortality of the higher human faculties] vanisheth."[24] Finding various difficulties in the notion of a divinely-infused soul, Overton maintained that plants and animals were given the power at the creation to "procreate their own kinde," whether vegetative, sensitive or rational, "without any transcendency of nature."[25]

Epicurean arguments against the incorporeality and immortality of the soul did not make an appearance in Overton's tract, and, as J.R. Jacob has pointed out, non-Epicurean forms of "pagan naturalism," with deistic and mortalistic implications were in circulation in the second half of the seventeenth century.[26] Hobbes, returning from Paris, did however base his philosophy—and notoriously his politics—on a Gassendist combination of a materialistic ontology and an epistemology of "appearances." "The *Universe* ... is Corporeall, that is to say, Body; and hath the dimensions of Magnitude, namely, Length, Bredth and Depth: also every part of Body, is likewise Body, and hath the like dimensions; and consequently every part of the Universe, is Body and that which is not Body, is no part of the Universe. ...Spirits have dimensions and are therefore really *Bodies*."[27] The subject of philosophy is, accordingly, body, its alterations, and appearances. and God is not treated by philosophy, insofar as philosophy deals with what is "capable of composition and resolution," whereas God is "Eternal, Ingenerable, Incomprehensible," and generally immune

[22]R.H. Kargon, *Atomism in England from Hariot to Newton*, Oxford: Clarendon Press, 1966, p. 7 ff.

[23]*The World Turned Upside Down*, London, Temple Smith, 1972; and *Milton and the English Revolution*, London: Faber and Faber, 1977.

[24]Richard Overton, *Man Wholly Mortal*, 2nd ed., 1655, p. 5.

[25]Ibid., p. 103.

[26]J.R. Jacob, "Boyle's Atomism and the Restoration Assault on Pagan Naturalism," *Social Studies of Science*, 8 (1978), pp. 211–233.

[27]Thomas Hobbes, *Leviathan*, ed. Richard Tuck, Cambridge, Cambridge University Press, 1996, Pt. IV. Ch. 46, p. 463.

to division and generation.[28] The Greek (Platonic-Aristotelian) preoccupation with incorporeals, Hobbes maintained, vitiated their understanding of natural and civil philosophy. "The Enemy . . .. has sown the tares of Spirituall Errors . . . by introducing the Daemonology of the Heathen Poets. That is to say, their fabulous Doctrine concerning Daemons, which are but Idols or Phantasms of the braine, without any reall nature of their own, distinct from humane fancy; such as are dead mens Ghosts, and Fairies. Third by mixing with the Scripture divers reliques of the Religion, and much of the vain and erroneous Philosophy of the Greeks, especially of Aristotle."[29]

Hobbes's claim that the whole universe was exclusively corporeal was an extreme view in 1651 and remained so, but, as Gassendi's neo-Epicurean system gained currency, especially after the posthumous publication of his *Syntagma* of Epicurean philosophy (1659), the possible fictivity of all incorporeal substances not resolvable into corpuscular constituents was a thesis that was increasingly discussed and debated. The philosopher who was instrumental in transmitting Gassendi's cleaned-up version of libertine Parisian materialism into England, avoiding Hobbesian exaggeration, was Walter Charleton, to whose case I now turn. For Charleton, the immortality of the soul was the "the grand Base of Religion", the keystone of its arch. "For if the Soul be mortal & subject to utter dissolution with the body: to what purpose doth all Piety and religion serve? What issue can we expect of all our prayers, of all our Adorations, of all our Self-denying acts of obedience, of all our unjust sufferings? Why should we worship God at all? Nay more, why should we consider whether there be a God or no?[30]

Charleton, who was born in 1620,[31] was a pupil of John Wilkins in Logic and Philosophy at Magdalen Hall (Hertford College) Oxford, later one of the King's Physicians, (both Ch. I and Char II) and a follower of Harvey, whose works on circulation and generation he knew. His first interests were anatomy, natural history and iatrochemistry. After a Helmontian phase, issuing in his *Ternary of Paradoxes* (1650), he discovered French philosophy, meeting Gassendi in 1653–1654, and became the chief conduit of Epicureanism in both natural and moral philosophy, though his *Physiologia Epicuro-Gassendo-Charletoniana* (1654), an important source for Newton and Boyle.[32] He compiled a *Natural History of the Passions* (1674); wrote on various topics in medicine, natural history, and the psychology of the emotions, and held various posts, including that of Harveian Librarian in his 86th year.[33] He died "in very poor Circumstances" at the age of 87.

---

[28]Hobbes, *Elements of Philosophy*, (1656) Pt I, Ch. 1, Sect. 8.

[29]*Leviathan*, Pt IV, Ch. 44, p. 418.

[30]Walter Charleton, *Immortality of the Human Soul*, p. 59.

[31]The secondary literature on Charleton is scant. The best source is Sabina Fleitman, *Walter Charleton (1620–1707) "Virtuoso." Leben und Werk*, Frankfurt and Berlin, Peter Lang, 1986. See also Lindsay Sharp, "Walter Charleton," *Annals of Science* 30 (1973) pp. 311–340.

[32]Kargon noted that Boyle's *History of Fluidity and Firmness* is taken practically verbatim from Charleton's *Physiologia*, both of which are indebted to Gassendi's animadversions on the tenth book of Diogenes Laertius. "Walter Charleton, Robert Boyle, and the Acceptance of Epicurean Atomism," *Isis* 55 (1964) pp. 184–192.

[33]Following Fleitmann, p. 8.

Charleton first attempted to defend the immortality and immateriality of the soul along unoriginal Cartesian lines in the *Darknes of Atheism* of 1652. His post-*Physiologia Immortality of the Human Soul* of 1657 turned away from a priori proofs and showed a real engagement with the arguments for the mortality of the human person, as well as considerable ambivalence about the possibility of actually knowing anything about this subject.

The *Immortality* takes the form of a dialogue between "Lucretius," representing the friend of Charleton's youth, John Evelyn, an admirer of Lucretius who later repudiated his earlier enthusiasm, and "Athanasius," the author of an unpopular treatise in natural philosophy corresponding to the *Physiologia*. Athanasius presents to Lucretius a panoply of arguments for the immateriality and immortality of the soul, which Lucretius refutes, and Lucretius presents arguments for its materiality and mortality which are refuted by Athanasius. The dialogue ends officially with a victory for Athanasius.[34]

Athanasius and Lucretius both avow themselves admirers of Nature. Athanasius is an experimental Philosopher, fresh from Oxford, and they agree that "who so enquires into the operations of Nature, by no other light than that of Books and solitary speculations, shall in the end find his head full of specious Termes, but empty of true and solid Science."[35] Lucretius, unlike his forebear, describes himself as believing in the immortality of the soul "as firmly as you, or any person living can", though desirous of arguments "as might for ever silence all Doubts and Contradictions and make a convert of my old master Epicurus." I am, he says, "an Epicurean in many things concerning Bodies; yet, as a Christian, I detest and utterly renounce the doctrine of that Sect, concerning Mens Souls."[36] He has laid on the "the disguise of a Contrary opinion . . .only to experiment the strength of your Allegations."[37] Athanasius, for his part, notes that Christians divide into two kinds: those whose faith is so powerful that they have no need for and no use for the assistance of the Reason. The other group, while not lacking in faith, "yet are glad when they can bring up the Forces of their Reason to assist them in the conquest of their fleshly oppositions."[38] It is to this latter group that Athanasius belongs.

Lucretius points out that the Doctors of the Church and the school-men have confessed that their arguments on behalf of the human soul are "not rigorously convincing, or such as constrain assert as inevitably as *Mathematical Demonstrations*."[39] Athanasius demurs mildly; some doctors, he says, "stiffly maintained" that their demonstrations were apodictical, and in any case, it is possible to provide reasons,

[34]The reader is reminded of Hume's *Dialogues*, at the conclusion of which, after Philo's devastating attack on natural theology, it is cheerfully agreed by all that the theist Cleanthes has the upper hand. As with other examples of theological dialogue going back to Lucilio Vanini, one has the impression that if the author really wanted to suffocate an objectionable doctrine, he would not have allowed its spokesperson so many pages.

[35]Charleton, *Immortality*, p. 5.

[36]Ibid., p. 185.

[37]Ibid., p. 152.

[38]Ibid., p. 57.

[39]Ibid., p. 61.

which though not equal in force to Geometrical Demonstrations, are such "as import either a Physicall or Moral evidence, sufficient to perswade a mind well affected toward truth, and free from the obstruction of prejudice." As long as they are more clear and certain than proofs that have been urged to the contrary, when added to holy writ they are "ineluctable."[40]

Athanasius's arguments for the immortality of the soul are borrowed from Gassendi and divide into the "physical" and the "moral." The physical arguments are drawn from consideration of the operations of the soul, its freedom and its ability to will what is good and honourable, preferring it to what is delicious and sensible, and its intellectual power to conceptualize without imagination, to think reflexively on itself, and to consider universals. The soul can take everything as her object, Athanasius says, both corporeal and incorporeal things.

These arguments from the nobility and excellence of the soul and its particular intentional relationship to the good and to its immateriality are familiar from Plato, Cicero and Descartes. They do not however seem to have a direct bearing on the question of *immortality*. Athanasius seems resigned to showing that if the soul can be shown to be incorporeal, contrary to what Lucretius maintains, the battle is largely won. He cites, the arguments familiar from Plato's *Phaedo*; whatever is immaterial has no parts, hence is not dissoluble, and hence "must of perfect necessity alwayes continue to be what it is."[41]

Lucretius insists that "substances Immaterial" are an absurdity or "somewhat too sublime for the comprehension of so humble and short-sighted a reason as mine is."[42] He cites Pomponazzi and Kenelm Digby (though not Hobbes!) to the effect that that "cognition is made by the working of our Phansy," demanding, "some more pregnant testimony, of the intellects knowing, without the immediate help of Images, pre-admitted by the Senses. " This argument continues, with Athanasius attempting to show that powers of the human soul such as self-awareness and cognition require its incorporeality and Lucretius insisting that all knowledge is a form of sensory representation, which can be referred to the corporeal animal soul.

It is the "moral" arguments that now have to bear most of the weight of the claim for immortality. These arguments comprise the universal consent of all men of all ages, nations, and religions to the doctrine of an afterlife, the appetite of immortality "naturally inherent in all men," and the necessity of divine justice. Yet here too Lucretius seems to gain the upper hand, for he points out that belief in immortality is not universal—Epicurus for one did not hold it—and that foolish opinions are common among men. Athanasius concedes that "erroneous Conceits many times spread themselves abroad, and diffuse by what subtle contagion I know not; especially when they have first been taken up upon presumption of Authority, Antiquity, Utility, and the like Inducements to belief. But as for such things, of which Nature

[40]Ibid., p. 62.

[41]Ibid., pp. 78–79.

[42]Ibid., p. 84.

herself hath implanted as certain Knowledge in our Minds; it is not vulgar for men to be mistaken in them." Rather than decaying like other superstitions, he says, this one "grows every day more strong and lively."[43]

Lucretius finally introduces the suggestion, already bruited in antiquity, that the immortality of the soul, like all religious doctrine, is a deception of priests, "it being at all times true, that such audacious Malefactors, as are not moved by the whole arme of the Civil Magistrate, will yet tremble at the finger of Divinity."[44] Is it not possible, Lucretius asks "that Men, casting about for various devices and imaginations to palliate and sweeten the sowernesse of the their Miseries, in this life, may have both invented this comfortable opinion of a state of future Immortality; and introduced the supposition of this provident justice of God; relating only to mens actions on purpose to support it?"[45]

The possibility that the immortality of the soul is an ideological doctrine—echoing the many contemporary attacks on mystifying priestcraft in the Interregnum—is expounded at some length. But Athanasius rejects the possibility that it is a "meer politique Fiction" on the grounds that "the first Law-Makers we read of in history;. . .found this Tenent of the Soul's Immortality settled and radicated in the hearts of the people, from the beginning of Mankind."[46] He admits the belief might have been manipulated by them, but denies that it is instilled by the powerful.

Lucretius finally concludes that Athanasius has satisfied all his doubts and solved all his objections about the immortality of the soul. "Yet whether you have so Demonstrated it, as to exclude all Dubiosity, and compell assent. . . in a pure Natural Philosopher" has to be left to the judge, Isodiocastes, to decide.[47] Isodiocastes insists that demonstration *more geometrico* is, of all forms, the most convincing and scientificall, but that for certain Metaphysical and even physical matters, "we ought not to require absolute Demonstrations." Yet the book seems to advertise its own failure to prove the case for the immortality of the soul against the atomist alternative and the political account of the origins of religious belief. Charleton's Dedicatory Epistle noted that "among my Readers, many may chance to dislike the book it self;" though they cannot fail to approve the Dedication. For, he admitted, his reasonings

> have not attained to that perfection and exquisite Rigour, as to satisfie those immoderately Curious Wits of our Age, who think it much beneath them, to acquiesce in any other Evidence but that of Demonstrations Geometrical (of which notwithstanding, the Argument of these my Discourses is absolutely uncapable.)[48]

---

[43] Ibid., p. 131.
[44] Ibid.
[45] Ibid., p. 147.
[46] Ibid., p. 132.
[47] Ibid., p. 186.
[48] Ibid., Epistle Dedicatory.

## IV Boyle, Bentley and Layton on Thinking Matter

It is as difficult to establish Charleton's true position on the existence of an incorporeal soul as it is Gassendi's. Charleton, by profession a physician, went on to publish *Epicurus's Morals*, in which he maintained that the greatest evils to befall man concern his body. Whatever his degree of belief in an incorporeal, immortal soul, he was, like most of his sober contemporaries, fearful of the social consequences of a thorough or widespread acceptance of mortalism. Mortalists and proponents of thinking matter, amongst them philosophers associated with "Socinianism" like John Milton,[49] took care to distinguish their position from Atheism. Blasphemy, denying or raising questions about God's existence, was still a hanging crime. Even Richard Overton had insisted that he was not questioning the resurrection; indeed, he said, an advantage of not supposing the soul to be naturally immortal is that there is room for preferential salvation of the virtuous. Thus the Epicurean blasphemy "Let us eate, and drinke, for tomorrow we die" is, he says, whether ingenuously or disingenuously, avoided.[50] Mortalists allowed that even if the human soul was corporeal, or an epiphenomenon of material organization, and not intrinsically immortal, God could revive or reassemble persons by fiat, or they could be made to reappear by some process of palingenesis.

Robert Boyle, who had eliminated forms from his ontology and reduced qualities to dispositions, struggled with the question of the human soul, admitting the radical inadequacy of the Cartesian arguments. He hoped desperately to find experimental evidence for the resurrection and also for witchcraft and demonic activity,[51] and he endowed the Boyle Lectures to continue the struggle against atheism, materialism, and mortalism. Richard Bentley, the first of the Lecturers, whose chief aim seems to have been to preserve Newtonian science from charges of atheism, preached a sermon on April 4, 1692, entitled *A Confutation of Atheism from the Faculties of the Soul, alias Matter and Motion cannot think*. His position was in turn challenged by Henry Layton (1622–1705) who brought the skeptical charge of "not proved," appealed to the voluntaristic conception of God as perfectly omnipotent and not limited by human reason, and, at the same time, advanced his own views on the material substrate of thinking.

> [W]ith God . . .all things are possible: whence it seems possible with him to make cogitative Body, notwithstanding our Preachers assurance to the contrary. . .Our Lord himself tells us that, that *with God all things are possible*; and it seems somewhat an odd Assertion that the Incapacity of Matter should hinder God from making what he will out of it; whereas

[49] On Milton and Mortalism, see Christopher Hill, *Milton and the English Revolution*, Ch. 25 et *passim*.

[50] Overton, *Mans Mortalitie*, pp. 106–107.

[51] See Michael Hunter's contribution in *La philosophie de Robert Boyle*, ed. Charles Ramond and Myriam Dennehy, Paris, Vrin 2009, (forthcoming)

> it seems rather, that he who made Matter out of nothing, can make anything out of Matter, and many other things than Men can imagine. . . ."[52]

Thomas Willis, the anatomist of the brain has dismissed with scorn, Layton said, the Cartesian hypothesis of animal insensibility.[53] Although we do not "thorowly know the Substance and Operation of the brain, nor the *ubi* or, *quomodo* such things are wrought in it" we can agree with Melancthon that the brain is the "seat and officina of cogitation."[54]

Men are not "bare machines," Layton said, for they have Sense and Reason. Like the beasts, however, they are made of matter "pulverized and rarified into the tenuity of a Cloud, impregnated with steams and juices no less, but more fine and active than the Vegetable Souls or Spirits of Plants . . .. I conceive Matter, thus constituted, impregnated, irrigated, and enlived, may, by Divine Power, be made cogitative."[55] Human art, he pointed out, can "communicate to dead and hard Matter a fixed and regular Motion: Witness *Architas* his Dove, and *Regiomontanus* his Eagle . . . Men would not then believe [this] could be done without an Immaterial Spirit."[56] And just as "those Atoms which we call Motes in the Sun . . .. of their own nature maintain a perpetual motion," it "seems a Compositum of such active Particles impregnated with rorid Steams and Juices apt to ascend by adhering to any solid Body, are not apt alone for Motion, but. . .by the hand and skill of a Divine Architect there may be made of such like Ingredients, a Cogitative Matter."[57]

## V Locke's Argument from Ignorance and his "Ideas"

Locke's *Essay* of 1689, and the various changes introduced into the second edition of 1694 and subsequent editions, needs to be interpreted against the background of Continental and English anti-Cartesian Gassendist skepticism.[58] Despite his earlier waverings, by 1683 Locke's reflections on personal identity had, according to Michael Ayers, veered away from his early mind-body dualism, and assumed "a starting point more favourable to the materialists."[59] The "thinking matter" proposal added in the second edition of 1694 is similar to Layton's in suggesting that we can't know that God has not "given to some Systems of Matter fitly disposed, a

---

[52]Henry Layton, *Observations upon a Sermon Intitulated,* A Confutation of Atheism from the Faculties of the Soul, *alias,* Matter and Motion cannot Think*: Preached April 4, 1692-By way of Refuation,* London 1692, p. 16, p. 18.

[53]Layton, *Observations*, p. 4.

[54]Ibid., p. 18.

[55]Ibid., p. 13.

[56]Ibid., p. 4.

[57]Ibid., p. 7.

[58]See, on the continental side, Thomas M. Lennon, *The Battle of the Gods and Giants; The Legacies of Descartes and Gassendi, 1655–1715*, Princeton, Princeton University Press, 1993.

[59]Michael Ayers, *Locke*, 2 vols., Routledge, London and New York, 1991, II: 255.

power to perceive and think," rather than joining to our bodies a thinking, immaterial substance, in virtue of which suitably organized systems might exhibit not only vegetable life, but also animal perception, and human reason.[60]

The proposal prompted the attacks of the longtime scourge of Epicureanism, Edward Stillingfleet, the Bishop of Worcester.[61] Many of Locke's critics were convinced, on the basis of the *Essay*, that he was a "Hobbist." John Edwards denounced him from the pulpit in a sermon at Cambridge in July 1699:

> It is *probable*, that as in *other* very considerable Points, so *here* he Symbolizes with the Philosopher of *Malmsbury*, in whose steps he affects to tread, and borrows some of his Thoughts. For . . .he follows him in his Opinion of the *Necessity* of only *One Fundamental Article of Christian Faith*, and in his notion of *Thinking matter*, and particularly in that of the likelyhood or possibility of the *Materiality of Humane Souls*, and of their tendency (on that Principle) to *Mortality*, and in his *Contempt of some parts of the Holy Scripture*, and in his avow'd disbelief of the *Resurrection of the same Body*. . .and in his general favouring of *Scepticism* and *Infidelity*, and his denial of *Natural* and *Inbred Notions*.[62]

Leibniz, in turn, appeared to consider Locke a Gassendist who held persons to be purely material machines.[63]

Locke was not an Epicurean in either the sphere of moral theory or of natural philosophy. As Ayers very appropriately remarks, Locke wanted rather to "demonstrate that scepticism about that which thinks in us is compatible with a belief in immortality."[64] At times, he even sounded disconcertingly open-minded with respect to the truth of Cartesian dualism. It is, he said, "as rational to affirm there is no Body, because we have no clear and distinct *Idea* of the *Substance* of Matter, as to say there is no spirit, because we have no clear and distinct *Idea* of the *Substance* of a Spirit."[65]

> [B]y putting together the *Ideas* of Thinking, and Willing, or the Power of moving or quieting corporeal Motion, joined to Substance, of which we have no distinct *Idea*, we have the *Idea* of an immaterial Spirit; and by putting together the *Ideas* of coherent solid parts, and a power of being moved, joined with Substance, of which likewise we have no positive *Idea*, we have the *Idea* of Matter. The one is as clear and distinct an *Idea*, as the other.... It is for want of reflection, that we are apt to think, that our Senses show us nothing but material things. Every act of sensation, when duly considered, gives us an equal view of both parts of nature, the Corporeal and Spiritual. For whilst I know, by seeing or hearing, *etc.* that there is some Corporeal Being without me, the Object of that sensation, I do more certainly know,

---

[60] Locke, *Essay* IV.III.6.

[61] Stillingfleet dealt extensive with Epicureanism in his *Origines Sacrae* of 1662.

[62] John Edwards, *The Eternal and Intrinsick Reasons of Good and Evil*, London, 1699, p. 28

[63] In Leibniz's *New Essays*, the author's spokesperson, Théophile, says, "The author [Locke] is pretty much in agreement with M. Gassendi's system which is fundamentally that of Democritus: he supports vacuum and atoms, he believes that matter could think, that there are no innate ideas, that our mind is a tabula rasa, and that we do not think all the time; and he seems to agree with most of M. Gassendi's objections against M. Descartes (NE 1.1, A VI vi, 70); cf. Locke, *Essay*, 1.iii.14.

[64] Ayers, *Locke*, II: 205.

[65] Locke, *Essay* II, 23.5.

> that there is some Spiritual Being within me, that sees and hears. This I must be convinced cannot be the action of bare insensible matter; nor ever could be without an immaterial thinking Being.[66]

But it is important to notice that Locke was not claiming in this context that an incorporeal soul existed and was known to exist. The term "Spiritual Being," referring to the thing he claimed to "know" to exist, was not the customary term of English dualists, and here refers to whatever thing has the power to see, hear, etc. "Bare, insensible matter," Locke thought, cannot be conceived to see, hear, etc. Nevertheless, as he explained later, we do not *know*, from contemplation of our own ideas, that is to say, of the appearances, that God has not superadded powers to bare insensible matter, in which case the Spiritual Being of which I am aware is just my own purely material, thinking self.

With his double negation, Locke scrupulously avoided affirming ontological doctrines about what really exists. Denying that his aim was to "anyway lessen the belief of the Soul's Immateriality" he emphasized that "it becomes the Modesty of Philosophy not to pronounce Magisterially, where we want that Evidence that can produce Knowledge"[67] He invoked the Gassendist principle that we have access only to appearances, now theorized by reference to a theory of "ideas," which are either composite "mixed modes" put together by the mind or else originating directly in matter, to wit., by contact with corporeal objects composed of particles possessing only primary properties.[68] The language of ideas evokes the spirit of Cartesianism, and even surrounds the *Essay* with an aura of spirituality. This is misleading. Ideas are more appropriately read as Gassendist appearances internalized; remember that Hobbes typically begins his expositions of Man not with Lucretian material first principles but with sensory appearances and the imagination.[69]

Locke proposed to consider morality as an appearance, to consider it as humans thought about it and experienced it, and a careful reading of the *Essay* that takes note of Locke's most extensive and labored discussions indicates that his intentions in writing it were to show that religious belief and moral practice could be maintained and defended even if God and an incorporeal, immortal soul were fictions, and even if men were motivated not by a Platonic orientation to the good, but merely by a desire for pleasure and an aversion to pain; the "unease" resulting from our various thoughts and perceptions. The key elements of this programme were: (a) an emphasis on the role of the *ideas* of an afterlife and an omniscient judge in strengthening moral resolve and enhancing moral conduct; (b) the elaboration of a phenomenological theory of personal identity that did not required a substantial and incorporeal soul to serve as a target of divine reward and punishment. Even if we are hedonistic machines, Locke thought, we are machines with the ability to act for our long-term happiness, once we acquire the relevant ideas.

---

[66] Locke, *Essay* II.23.15.

[67] Ibid.

[68] Locke, *Essay* II.i.

[69] Note the strictly phenomenological opening of *Leviathan*, Ch. 1.

Locke harped in Book IV of his *Essay* on the impossibility of our understanding how atoms might produce sensory experiences, and how atomic configurations and motion might explain the powers and appearances of substances, indeed on the impossibility of knowing that there are atoms. "[A]s to a perfect *Science* of natural Bodies, (not to mention spiritual Beings,) we are, I think, so far from being capable of any such thing, that I conclude it lost labour to seek after it."[70] His sacrifice of the ideal of science as knowledge of the inner nature of corporeal substances was not however very heavy in his eyes, and with it—as some of the "rubbish" he intended to clear away—went spiritual substances and their supposedly innate ideas, ideas not caused by contact with corporeal things. Because all sensory experiences were, in his view, caused by corporeal objects, and because all "ideas of reflection" and mixed modes presupposed sensory experiences, the existence of matter and its causal powers, unknowable though it was, was consistent with his epistemology, his psychology, and his moral theory, while the unknowable truth about incorporeal entities was logically irrelevant to both. By giving out, on the epistemological level of his discourse, that the existence of incorporeal human souls was no more doubtful in his eyes than the existence of matter, Locke was able to construct a system that in fact privileged matter, absolutely requiring it as an explanatory principle. By locating both unsensed corporeal things and unsensed incorporeal things outside the realm of practical, efficacious human knowledge, Locke was trying to articulate the forms of new sciences—natural and moral—that would require no reference to unexperienced entities. He could now sketch a conception of scientific and moral knowledge based on Epicurean-Gassendist premises, corresponding to what he took to be the actual practice of his heroes, especially Boyle in chemistry, Newton in applied mathematics, and Sydenham in medicine. His critical stance enabled him to elude charges of materialism—though vigilant critics like Edwards and Leibniz professed to see through his skepticism to his supposed Hobbist commitments.

## VI The Science of "Mere Corporeal Nature"

The shape of seventeenth century philosophy and epistemology were determined by the revival of an ontology and an epistemology that advanced the exclusively material principles discredited by Plato and Aristotle, who had introduced incorporeal entities and principles welcomed by Christian theologians and moralists alike. This earlier ideal of science—appearance or idea-based in epistemology, atom-based in ontology, careless of logical relations but wedded to mechanical accounts—was the object of recovery efforts that were carried to new heights of sophistication and philosophical elaboration. Material entities were not only representable to the human visual imagination, but susceptible to manipulation and control. Where Aristotelian *scientia* was famously contemplative and personally gratifying to the individual who had secured it, early modern experimental science presupposed the

[70]Locke, *Essay*, IV.iii.29.

group. It was diffused and often confused, but also ambitious and meddlesome. Hylomorphs, with their own indwelling principles of control, had to be ascribed their own nonhuman agendas, but merely material things were passive in the hands of men, if only (this proved to be not so easy) the right procedures and apparatus could be developed for accessing the tiny particles of which they were composed. By rejecting the Aristotelian premise that the true physicist studies matter and form, and that the study of the soul falls within the science of nature, and by offering to turn over the discourse of the soul to the theologians, the experimental philosophy pretended to apologize for its weakness but in fact sought to strengthen its position. Its subject was, to employ Boyle's term, "mere corporeal nature." Meanwhile, empirically-minded Christians shifted their attention from attempted proofs and demonstrations of the existence of unexperienced states and entities to the provision of pragmatic reasons for belief in them. Both Boyle and Bentley suggest that pragmatic, life-sweetening belief in God and the future state in what William James would call a situation of forced and momentous choice is well-warranted.

"Now as long as all men desire to know as much as possible either according to experience or within the limitations of appearances," Gassendi had said in his *Exercitationes*, "it is true that they desire to know these things at nature's promptings: but as soon as they claim to know other things beyond these; either inner natures or necessary causes, that is the sort of knowledge that belongs to angelic natures, or even to the divinity, and is not proper for paltry men."[71] It would be a mistake to suppose that the epistemological quietism of Gassendi and Locke left everything just as it was. The famous sceptic who professed to be concerned only with the appearances boasted at the same time, "I give a soul to semen, I restore reason to animals; I find no distinction between the understanding and the imagination."[72] These were prophetic remarks.

---

[71] Gassendi, *Exercitationes*, tr. Brush, p. 104.

[72] Ibid., p. 22.

# *Scientia* and *Inductio Scientifica* in the *Logica Hamburgensis* of Joachim Jungius

**Stephen Clucas**

In a recent article which seeks to re-examine the development of the Cartesian method in the context of late sixteenth- and early seventeenth-century debates about methodology, Timothy Reiss re-examined the influential thesis of J. H. Randall's *The School of Padua and the emergence of modern science* (1961) which presented the logical writings of Jacopo Zabarella as an important stage in the development of scientific method in the seventeenth century.[1] Reiss makes important qualifications to Randall's thesis and in particular he shows how Zabarella neglected a set of concerns which came to assume particular prominence in later conceptions of scientific method, the absence of which make him a questionable candidate for a founder of modern scientific method. The restricted and subordinate role played by experience and observation in Zabarella's concept of the scientific method (in spite of Randall's claims to the contrary)[2] and the correspondingly supervenient role which he gives to the *intellectus agens* leads Reiss to conclude that Zabarella's method is "far from anything we could recognise today as 'scientific method'", and has besides "nothing

S. Clucas (✉)
Birkbeck, University of London
e-mail: s.clucas@bbk.ac.uk

[1] Timothy J. Reiss, "Neo-Aristotle and method: Between Zabarella and Descartes" in Stephen Gaukroger, John Schuster and John Sutton (eds.) *Descartes's Natural Philosophy* (London and New York: Routledge, 2000), pp. 195–227. John Herman Randall, Jr, *The School of Padua and the emergence of modern science* (Padua: Editore Antenore, 1961). The first chapter, where Randall deals with Zabarella's concept of method and its Aristotelian antecedents was first published (without Latin texts) in the *Journal of the History of Ideas*, 1 (1940), 177–206. For more recent studies of Zabarella and scientific method see Antonino Poppi, *La Dottrina della Scienza in Giacomo Zabarella* (Padua: Editrice Antenore, 1972), esp. Ch. 5, "Analisi del metodo scientifico", pp. 161–195 and Ch. 6, "La dottrina della scienza", pp. 197–294 and Heikki Mikkeli, *An Aristotelian Response to Renaissance Humanism: Jacopo Zabarella on the Nature of the Arts and Sciences* (Helsinki: Societas Historica Finlandiae, 1992), esp. Ch. 3, "Logic as an Instrument of Science", pp. 21–44.

[2] Randall claims that it was the "originality of Zabarella ... thus to set off a 'scientific experience' from mere ordinary observation, the accidental or planless collection of particular cases." (Randall, 55), but he does not present any texts to substantiate this claim.

T. Sorell et al. (eds.), *Scientia in Early Modern Philosophy*, Studies in History and Philosophy of Science 24, DOI 10.1007/978-90-481-3077-1_4,

directly in common with a 'geometrical method'".[3] Crucially, Reiss argues, "The relation between experience, perception of a material history and mathematical rule that others were seeking and forging remained foreign to Zabarella."[4] In this paper, I will be looking at another Aristotelian figure of the early seventeenth century, an older contemporary of Descartes, who made a significant if relatively neglected contribution to the debate on scientific method, who was also "seeking and forging" a relationship between experience, mathematics and logic as a way of providing certain, demonstrative knowledge in natural philosophy.

Joachim Jungius was born in Lubeck in 1587, and educated in the Universities of Rostock and Giessen. In 1609 he became Professor of mathematics at Giessen. In 1616 he studied medicine at Rostock, and continued his studies in Padua under Cesare Cremonini. He was made doctor of medicine in 1619, and practised medicine in Rostock before becoming Professor of mathematics in 1624.[5] In 1622 he founded a short-lived "Societas Ereneutica", which encouraged communication between empirically-minded scientists.[6] From 1629 to 1657 he was rector of the Akademisches Gymnasium and (until 1640) of the Johanneum, both in Hamburg. It was while he was in Hamburg that Jungius wrote his *Logica Hamburgensis*—published in 1638—which, as its title-page announced, was "written for use in the Hamburg schools" (*in usum Scholae Hamburgensis conscriptae*).[7] The *Logica* was, however, far from a conventional pedagogical textbook, and its fourth book which was entitled *Logicae Specialis* dealt with the application of logical induction to the natural sciences in a novel and controversial fashion. Local reactions to the book were extremely hostile and Johann Scharff (1595–1660), Professor of Practical Philosophy and Extraordinary Professor of Theology at the University of Wittenberg felt

[3] Reiss, "Zabarella and Descartes", pp. 208–209. This criticism of Randall's claims for Zabarella's method by appeal to Zabarella's use of the *intellectus agens* was also raised by H. Skulsky, "Paduan Epistemology and the Doctrine of One Mind", *Journal of the History of Philosophy*, 24 (1968), 231–40.

[4] Reiss, "Zabarella and Descartes", p. 210

[5] Gottschalk Eduard Guhrauer, *Joachim Jungius und sein Zeitalter. Von G. G. Guhrauer. Nebst Goethe's Fragmenten über Jungius* (Stuttgart and Tübingen: J. G. Cotta'scher, 1850), pp. 1–79.

[6] See Guhrauer, *Joachim Jungius und sein Zeitalter*, pp. 69–71, and Donald R. Dickson, *The Tessera of Antilia: Utopian Brotherhoods & Secret Societies in the Early Seventeenth Century*, Brill Studies in Intellectual History, 88 (Leiden: Brill, 1998), "Jungius and the *Societas Ereneutica*", pp. 91–101.

[7] Joachim Jungius, *Logica Hamburgensis, hoc est, Institutiones Logicae in usum Scholae Hamburgensis conscriptae, & sex libris comprehensae autore Joachimo Jungio Phil. ac Med. D. Gymnasii ac Scholae Rectore. Libri tres priores Logicam Generalem complexi jam prodeunt* (Hamburg, 1638). For a modern critical edition of the Latin text with German translation and extensive annotations see Rudolf W. Meyer, *Joachimi Jungii Logica Hamburgensis* (Hamburg: J. J. Augustin, 1957). Meyer's work was later supplemented by Wilhelm Risse's compilation of additional material relevant to the *Logica*, and its contemporary reception. See Wilhelm Risse (ed.), *Logicae Hamburgensis additamenta. cum annotationibus edidit Wilhelm Risse*, Veröffentlichung der Joachim Jungius-Gesellschaft der Wissenschaften, Hamburg, 29 (Göttingen: Vandenhoeck and Ruprecht, 1977).

moved to respond. His *Lima logicae Hamburgensis qua errata Joachimi Jungii*, in which the "errors" in Jungius's logic are "emended and perfected" by referring them to the works of Aristotle "and other approved authors," was published in Wittenberg in 1639.[8] Although a great part of Scharff's criticism focused on the extent to which Jungius had "departed ... from the ancient and most excellent decrees of the Logicians" (*recedit ... ab antiquis et probatissimis Logicorum decretis*),[9] he was also deeply offended by Jungius's application of logic to the empirical investigation of nature. "The end of logical demonstration is certainly knowledge," Scharff conceded,

> but it is nothing like the ambiguous sense which Jungius makes of it, but rather that true knowledge which is necessary and evident in every way, and quite different from that which is found in the *Topics* concerned with probabilities and contingent things [...] Scientific method is not satisfied by *a dianoetic process from things which are better known* as the author Jungius has stated, but requires something higher and far more sublime. The principles of demonstration ought to be general, essential and universal [*kata pantos, kath'auto et katholou*] not particulars, accidentals or singular things.[10]

Outside of the narrowly Aristotelian milieu of the German universities, however, Jungius's work found great favour. Joachim Hübner, a collaborator of the Anglo-German Baconian reformer Samuel Hartlib, was a great supporter of Jungius's work. "A hundred logics like Scharff's", wrote Hübner, "do not have as much weight, I judge, as the fourth book of Jungius's *Logica* alone, where he gives himself more latitude, and the examples alone, which he subordinates to his precepts, show how remarkably focussed he is on the mathematical arts and natural science."[11] René

8 Johann Scharff, *Lima Logicae Hamburgensis qua errata Joachimi Jungii Med. Doctoris, et Gymnasii Hamburg. Rectoris, expoliuntur atque emendantur: Genuina vero logices praecepta ex Aristotele aliisque probatis autoribus conformantur exactius, adornata opera et studio Iohannis Scharfii* (Wittenberg, 1639). On the Scharff controversy see Risse, *Additamenta*, "Streitschriften mit Johannes Scharfius", pp. 311–378, which includes an edition of Scharff's work (pp. 323–378). For a summary of Scharff's career and publications see Charles Lohr, *Latin Aristotle Commentaries, II: Renaissance Authors*, Corpus philosophorum medii aevi. Unione accademica nazionale. Subsidia, 6 (Firenze: Leo Olschki, 1988), p. 409.

9 Scharff, *Lima Logicae Hamburgensis*, in Risse, *Additamenta*, p. 325.

10 Scharff, *Lima Logicae Hamburgensis*, in Risse, *Additamenta*, p. 327: "*Finis demonstrationis Logicae* est sane scientia, sed non quaevis, ut ambiguam hic facit vocem Jungius, sed vera illa, quae a Topicis, probabilibus, atque contingentibus longe distat, omnino evidens, et necessaria est. [...] Methodus scientifica non contenta est *processu dianoetico a notioribus*, ut autor Jungius quidem statuit, sed altius et longe sublimius quid exigit. Principia demonstrandi *kata pantos, kath'auto et katholou* sint oportet, non particularia, non accidentalia, non singularia." Scharff's three principles of demonstration are taken from Aristotle's *Posterior Analytics*, I, iv, 73a.

11 Copy letter to Jan Amos Comenius, 7 December 1640, Hartlib Papers, HP, 7\78\10A: "centum Logicae Scharffianis similes tanti ponderis meo judicio no[n] sint, quam unicus quartus liber Logicae Iungianae, ubi aliquanto plus sibi permisit, et sola exempla quae hinc inde praeceptis suis subjungit, quam egregie in mathematicis artibus et scientia naturali versatus sit, ostendunt [...]." The editor of Comenius's correspondence, Jan Kvačala, identifies the anonymous correspondent as Hübner. See Jan Kvačala, *Korrespondence Jana Amosa Komenského, Nová sbírka*, 2 vols (Prague: České akademie Císaře Františka Josefa, 1897–8), I, 102. All quotations from the

Descartes, having seen a copy of Jungius's *Logica*, is said to have described him as "a man of subtle intelligence" (*hominem subtilis ingenii*).[12] The encyclopaedist Jan Amos Comenius in his *Pansophici libri delineatio* enthusiastically compared Jungius's work to Bacon's *Novum Organum*:

> Bacon has uncovered not a little in his marvellous *Organum* concerning the infallible way of profoundly investigating the things of nature. Jungius from Saxony is now labouring to bring that art of deductive Logic to its perfection, so that the truth of propositions can be built up, [and] the deceptions of sophisms destroyed, with the same certainty with which any Euclidean problem may be apodictically deduced.[13]

In his history of early-modern logic, *Die Logik der Neuzeit*, Wilhelm Risse noted the high esteem which the logic of Joachim Jungius enjoyed in the late seventeenth century:

> Leibniz commended Jungius as one of the best logicians, far surpassing Descartes. He declared the *Logica Hamburgensis* to be the most important logic of the seventeenth century. This high praise is fully justified, provided one sees the fullest possible enumeration of defined rules and theorems as the criterion of logic. For Jungius drew together encyclopaedically all the resources of traditional theory more completely than any other logic of his time.[14]

Although Jungius's considerable reputation in seventeenth-century Europe is relatively well known in German history of science circles and has been the subject of a number of important studies,[15] his name is still obscure to Anglo-American historians despite the fact that one of his foremost German interpreters Hans Kangro

---

Hartlib Papers are from *The Hartlib papers: (CD-ROM) a complete text and image database of the papers of Samuel Hartlib (c. 1600–1662) held in Sheffield University Library, Sheffield, UK* (Sheffield: HROnline, Humanities Research Institute, University of Sheffield, 2002).

[12] Hartlib Papers, *Ephemerides*, 1640, 30/4/56A.

[13] J. A. Comenius, *Pansophici libri delineatio* in *J. A. Comenii Opera didactica omnia. Variis hucusque occasionibus scripta, diversisque locis edita: nunc autem non tantum in unum, ut simul sint, collecta, sed & ultimo conatu in Systema unum mechanicè constructum, redacta* (Amsterdam: Laurentius De Geer, 1659), p. 426: "Non est nihil, quod *Verulamius* mirabili suo organo rerum naturas intime scrutandi modum infallibilem detexit. Et *Jungius* Saxo Logicae nunc arti ad eam perfectionem deducendae insudans, ut propositionum veritas adstrui, sophismatum deceptiones destrui queant, ea certitudine qua *Euclideum* quodvis problema apodeicticè deducitur."

[14] Wilhelm Risse, *Die Logik der Neuzeit*, 2 vols, Geschichte der Logik (Stuttgart-Bad Cannstatt: Friedrich Fromann Verlag, 1964), I, 521–2: "Als einen der besten, Descartes weit übertreffenden Logiker hat Leibniz Jungius gerühmt. Die *Logica Hamburgensis* ist als bedeutendste Logik des 17 jahrhunderts bezeichnet worden. Dieses hohe Lob ist voll berechtig, sofern man den Wertmaβstab der Logik in der möglichst vollständigen Aufzählung formulierter Regeln und Sätze sieht. Denn Jungius faβt so vollständig wie kein anderer Logik seiner Zeit den gesamten Bestand des überlieferten Lehrgutes enzyklopädisch zusammen." For Leibniz's views on Jungius and his logic see Risse, *Additamenta*, pp. 379–384, and Meyer, 1957, p. ix.

[15] Hans Kangro, *Joachim Jungius' Experimente und Gedanken zur Begründung der Chemie als Wissenschaft. Ein Beitrag zur Geistesgeschichte des 17. Jahrhunderts* (Wiesbaden: Franz Steiner Verlag, 1968) and Christoph Meinel, *Die Bibliothek des Joachim Jungius: ein Beitrag zur Historia litteraria der frühen Neuzeit* (Göttingen: Vandenhoeck & Ruprecht, 1992).

wrote a substantial article on him for the *Dictionary of Scientific Biography*.[16] For this reason alone it is worth reflecting on the particular contribution which Jungius made to the development of scientific method in the seventeenth century, and in particular the specific way in which he bridges the gap between the Aristotelian idea of *scientia* as certainly demonstrative knowledge and a new orientation towards the empirical investigation of nature.

As Reiss suggests in his article, and as Peter Dear has argued in a series of important studies of Aristotelian natural philosophy, one of the most important steps in the development of scientific method hinged upon the role allotted to experience and experiential data in the investigation of nature. As Dear has shown, the place of experience in Aristotelian natural philosophy was strictly subordinate, and was most commonly appealed to for examples confirming a priori truths:

> Since the point of Aristotelian scientific demonstration was to derive conclusions deductively from premises that were already accepted as certain – as with those of Euclidean geometry – there was no question of testing the conclusions against experience. The proper place for experience was in grounding the inductive generalizations contained in the original premises [. . .]. Once they had been established, so too, in effect, had the conclusions deduced from them.[17]

As Reiss notes, Zabarella gave no special place in his philosophy to observation, which he saw merely as the transmission of simulacra of the knowledge present in the mind of God to the *intellectus passibilis* by means of the *intellectus agens*.[18] Aristotelian natural philosophers in the early seventeenth century, however, began to attend more closely to the distinctions between experience, phenomena and observation, and formulated a conception of observation as *regulated* experience, which could be legitimately appealed to in making truth-claims in natural philosophy.[19]

This is precisely the kind of argument which Jungius mounts in the fourth book of the *Logica Hamburgensis*. In the first five aphorisms of Chapter 4, *De experientia*, Jungius distinguishes between the various senses in which experience can be taken, ending with the sense which is most pertinent to the natural philosopher:

> 1. *Experience* is occasionally taken to mean an *accident of the senses*, that is an action, passion, change, or event concerning a thing which we investigate or undergo, and approach via the senses. It is otherwise called *Experiment* (πειρα) or *Phenomenon*, that is something appearing to the senses.
> 2. Secondly *experience* may be taken to mean *sensation* itself, or the sensitive perception of the Phenomenon.

---

[16] Charles Coulston Gillispie (ed.) *Dictionary of Scientific Biography*, 16 vols (New York: Scribner's, 1970–80), VII, 193–6. For further biographical information on Jungius see Guhrauer, *Joachim Jungius und sein Zeitalter* and Martin Fogel, *Memoriae Joachimi Jungii mathematici summi* (Hamburg, 1657).

[17] Peter J. Dear, "Narratives, Anecdotes, and Experiments: Turning Experience into Science in the Seventeenth Century", in Peter J. Dear (ed.) *The Literary Structure of Scientific Argument: Historical Studies* (Philadelphia: University of Pennsylvania Press, 1991), pp. 135–163 (p. 143).

[18] Reiss, p. 208.

[19] Reiss, p. 209. Cf. Peter Dear, *Discipline and Experience: The mathematical way in the Scientific Revolution* (Chicago and London: Chicago University Press, 1995), pp. 47–57.

3. Thirdly it signifies *the cognition of the propounding intellect*, arisen from the perception of particular things, or from immediate cognition, that is an object presenting itself so that it is present for the one apprehending. It is thus called *actual experience*.
4. Lastly it sometimes designates a *habitus* originating from the operation of actual experience.
5. In this place let us deal with the *actual experience*, which is the second activity of the intellect, that is to say an assertion [*enuntiatio*], and with respect to scientific Method, a *Proposition*. It is thus called an *empirical Proposition*, that is to say an *empirical Theorem*.[20]

He also distinguishes (in aphorisms 17–31) between common experience and that which comes via observation.[21] Common experience is:

acquired in passing when we are dealing with something else, such as *Water dissolves salt, Lead is melted by fire, a Pig is an animal which is multiparous*. For they are aware of these things who either deal with or observe natural things in passing [although] not for the sake of knowing something theoretically.[22]

Experience which comes through observation, on the other hand, is either an "intellectual experience" or "experience-through-observation", or (properly):

an ordered series of sense perceptions deliberately set up for the purpose of knowing; of this kind are, *The observations of Astronomers, Anatomists, Physicians, and likewise the Chemical observations carried out by a philosopher*. [23]

Experiences are either confused (*confusa*) or distinct (*distincta*)—common experience "is nearly always confused" whereas "that which becomes known through observation is sometimes confused and sometimes distinct, depending on whether

---

[20] Jungius, *Logica* 1638, pp. 291–292: "*Experientia* interdum pro *sensili accidente* accipitur, hoc est pro actione, passione, mutatione, eventu circa rem, quam exploramus, experimur, & sensui subjicimus. Dicitur alias *Experimentum* πειρα item *Phaenomenon*, hoc est aliquid sensibus apparens. 2. Secundo *experientia* sumitur pro *sensione* ipsa, sive sensitiva perceptione ipsius Phaenomeni. 3. Tertio *cognitionem intellectus enuntiativam* significat, ex rerum singularium perceptione ortam, sive ex cognitione aliqua praesentanea, hoc est objectum suum praesens ut praesens apprehendente. Dicitur etiam *actualis experientia*. 4. Denique *Habitum* quandoque designat ex actualis experientiae exercitio prognatum. 5. Hoc loco de *actuali experientia* agendum quae operatio est intellectus secunda, videlicet *enuntiatio*, & respectu scientificae Methodi *Propositio*. Dicitur etiam *Propositio empirica*, item *Theorema empiricum*."

[21] Jungius, *Logica*, 1638, IV, iv, 17, p. 293: "17. Experientia porro alia *vulgaris* est, alia *per Observationem orta*."

[22] Jungius, *Logica*, 1638, IV, iv, 18, p. 293. "*Vulgaris* Experientia ἐμπειρία ἡ τυχουσυα ea est, quae obiter nobis & aliud etiam agentibus acquiritur ut *Salem aqua liquescere, Plumbum igne fundi, Suem esse animal multiparum*. Haec enim norunt etiam ij, qui non sciendi causa res naturales vel tractant, vel obiter intuentur."

[23] Jungius, *Logica*, 1638, IV, iv, 20–21: pp. 293–294 "20. *Observatio priori modo*, hoc est, *proprie sumpta* est ordinata series sensionum certo consilio instituta sciendi gratia; cujusmodi sunt *observationes Astronomorum, Anatomicorum, Medicorum, item Chimicae observationes a philosopho institutae*. 21. *Observatio posteriori modo* intellecta est experientia sive enuntiatio, quam intellectus format observationi proprie dictae fidem habens, ac proinde rectius *Experientia-per-observationem* dicitur."

the observation is conducted with more or less diligence, industry and skill."[24] Common experience is confused either with respect to its parts (when we impute to something that which is true only of one or more of its parts), or with respect to accidents, when an accidental attribute is mistaken for an essential one (if someone were to believe, for instance, that butter caused wounds to sting, when it was the salt in the butter which caused the discomfort).[25] Distinct experiences are ones where the accidental and the essential, and the properties belonging to the part and the whole, are properly distinguished. This is brought about by regulation, that is to say the logical ordering of observations:

> Distinct experience comes about when Observations are established in order, such as when one *proceeds gradually from simple things to composites, from pure things to mixed things, from less composite things to more composite things*, and this order is maintained not only for *agents*, but also for patients.

Significantly, Jungius here sees logic not just as a way of deducing certain truths from already-established truths of nature, but as a way of organising the *investigation* of nature. "For example", he continues:

> *if anyone wishes to explore the nature and properties of oil, let him acquire an oil which is neither crude, nor immature, rancid nor salted, nor medicated, for example with rue or dill, and obviously neither hot nor cold, but tepid; then he will apply it to similar rather than to dissimilar bodies; because if one were busy exploring these properties in the human body, one would first put it to the test on a healthy and well-tempered body, rather than on a sick one, and first on a simple rather than a complex disease.*[26]

Jungius was renowned for his expertise as an experimental chemist, and his comparison of chemical investigation and medical diagnosis is a telling one given his medical training in Rostock and Padua.[27] It is no surprise then that he invokes Galen as an authority for observational method in aphorism 30:

---

[24] Jungius, *Logica*, 1638, IV, iv, 29, p. 296: "29. Experientia vulgaris fere semper confusa est: Quae vero per observationem innotescit modo confusa modo distincta est; prout minore vel majore diligentia, industria, arte observatio administratur."

[25] Jungius, *Logica*, 1638, IV, iv, 25–6, pp. 294–295.

[26] Jungius, *Logica*, 1638, IV, iv, 28, pp. 295–296: "28. Oritur autem distincta experientia, cum Observationes eo ordine instituuntur, ut *a simplicibus ad composita, a sinceris ad permixta, a minus compositis ad compositiora sensim procedatur*, isque ordo non solum in *agentibus* teneatur sed & in *patientibus*. V. g. *Si quis olei naturam ac vires explorare cupiat, is oleum sumet neque crudum sive immaturum, neque rancidum neque salitum, neque medicatum, puta rutaceum, anethinum, imo neque evidenter calidum aut frigidum, sed medio modo se habens; idipsum autem admovebit prius similaribus quam dissimilaribus corporibus; quod si vires illius in corpore humano explorare satagat, prius eas periclitabitur in corpore sano & bene temperato, quam in aegro, & prius in simplici, quam in composito morbo.*" For a discussion of Descartes's analogous use of "clear", "distinct" "simple" and "complex", see René Descartes, *A Discourse on the Method of Correctly Conducting One's Reason and Seeking Truth in the Sciences*, ed. and trans. Ian Maclean (Oxford: Oxford University Press, 2006), pp. l–li.

[27] On Jungius's chemical experiments see Kangro, 1968, cit. fn. 15 above.

> There is much to be gained from arranging observations clearly, as indeed Galen advises, the terms of propositions (the truth of which are ascertained by observation) have to be extricated from premised homonyms by means of notional definitions.[28]

The real surprise here, however, is the casual way in which Jungius introduces the idea that the truth of premises "should be ascertained by observation" (*veritas observatione exploranda est*). In this Jungius reverses the usual scholastic procedure whereby experiences (whether observational or not) would be used to confirm or corroborate the universal truths propounded in propositions rather than to establish those truths, and opens the way to a philosophy (which he called the "protonoetic philosophy", *philosophia protonoetica*) which would place a pronounced emphasis on experiment and observation as well as on logical induction.

This development of a logical method whose focus is on experience and observation, can be seen clearly in the eighth chapter of book four (*De Inductione scientiali*). Having dealt with induction in general (which deals with the necessary or the probable indifferently) in book three (in Chapter 22, *De Inductione*), he now turns to what he calls "epistemonic or scientific" induction (*scientialis* sive *epistemonica*) which deals only with necessary truths.[29] "It is peculiar to this form of Induction", Jungius says,

> that it consists of premises that *depend on experiences alone*, that is, known either through the senses, or through reflective attention of the intellect, or through another prior induction.[30]

Scientific induction then deals with what in Chapter 3 he calls "actual experiences" or "empirical Theorems".[31] Significantly here Jungius distinguishes his own treatment from that of Zabarella, although it seems clear that Zabarella's *regressus* had an enormous influence on his own conception of logic in its application to the natural sciences.[32] Zabarella, he says, has confused scientific induction, whose premises consist only of experiences, and demonstrative induction, whose premises are "truths known through demonstration".[33] The latter is "pure" induction while

---

[28] Jungius, *Logica*, 1638, IV, iv, 30, p. 296: "30. Ad observationes distinctè instituendas multum confert, monente etiam Galeno, terminos propositionum, quarum veritas observatione exploranda est, ab homonymijs per praemissas notionales definitiones habere extricatos."

[29] Jungius, *Logica*, 1638, III, IV, viii, 1, p. 313. His treatment of induction in general is III, xxii, pp. 237–245.

[30] Jungius, *Logica*, 1638, IV, viii, 2, p. 313: "Habet autem & hoc praeterea peculiare haec Inductio, quod sumptionibus constat *soli experientiae innixis*, hoc est vel per sensum, vel per animadversionem intellectus reflexivam, vel per inductionem aliam priorem cognitis."

[31] See fn. 20 above.

[32] On Jungius's simultaneous debt to and criticism of Zabarella see Meyer, 1957, xv-xvi. On Jungius's adoption of Zabarella's definition of logic as an instrumental discipline see Mikkeli, 1992, p. 58.

[33] Jungius, *Logica*, 1638, IV, viii, 3, p. 313: "3. Si enim sumptiones Inductionis alicujus necessario—verae per demonstrationem innotuerint, jam non *scientialis* aut *epistemonica*, sed *apodictica* sive *demonstrativa Inductio* dicitur, quae in doctrina Demonstrationis explicabitur." 4. Jungius, *Logica*, 1638, IV, viii, 4, p. 313: "Zabarella *scientialem* & *demonstrativam Inductionem*,

the former is "mixed scientific induction" which is compounded of simple notions (*dianoeas*) or notions derived from previous inductions, which he calls "subservient inductions" (*Inductioni inservientes*), and these are:

> partly *syllogisms*, of which the premises are either nominal definitions or experiences, or propositions concluded from nominal definitions and experiences; partly the immediate consequences derived from the continually remaining notions.[34]

Jungius gives an example of a mixed scientific induction from what he calls "the most straightforward science of all" (*scientia omnium facillima*), arithmetic. In aphorisms 9–20, he shows how the proposition "that two consecutive triangular numbers added together makes a squared number" can be proved by induction. Through a mixture of premises derived from nominal definitions and experiences (and Jungius considers particular acts of intellection—such as mathematical statements like "Every distinct notion of a triangle is compounded of notions of surface, point, line and the ternary" to be "inward experiences"),[35] he shows how the problem can be resolved into simple elements, and then by syllogistic proofs, compounded into a complex notion which proves the original statement, and others like it. The simple elements of scientific or epistemonic induction Jungius calls "prosyllogisms" (that is, syllogisms whose conclusions become the first premise of another syllogism). "These prosyllogisms", he says,

> depend in part on *nominal definitions* which, as we said above, are propositions which are, of necessity, not strictly accurate truths; and in large part they depend on particular *empirical propositions*.[36]

While Jungius seems here to concede that the logical investigation of nature does not have the status of *demonstrationes certae*, i.e. by conceding that the premises on which the induction is based are "not strictly accurate truths" (*non exquisita vera*), nonetheless it seemed to offer to Jungius the best way of proceeding in the natural sciences. In particular it offered a way to re-open questions which had been closed off by received opinion:

> The intellect is so swift in these things, that having been perceived beforehand with nominal definitions, that it does not require *clear and distinct forms of simple notions*; but rather is entangled and exhausted by these [forms], rather than assisted by them.[37]

---

ut aequipollentia usurpat: verum nos ea, quae reipsa discrepant, appellationibus quoque distinguenda ducimus."

[34] Jungius, Logica, 1638, IV, viii, 8, p. 314: "*Dianoeae* istae *Inductioni inservientes* sunt partim *syllogismi*, quorum sumptiones sunt vel nominales definitiones, vel experientiae, vel propositiones ex definitionibus nominalibus & experientijs conclusae; partim *immediatae consequentiae* dianoeis reliquis continuandis adscitae."

[35] Jungius, *Logica*, 1638, IV, iv, 15, p. 293.

[36] Jungius, *Logica*, 1638, IV, viii, 21, p. 317: "Pendent enim prosyllogisimi isti partim e *definitionibus nominalibus*, quae, ut supra dictum, propositiones sunt necessitate non exquisita verae; partim ex *empiricis propositionibus* admodum specialibus."

[37] Jungius, *Logica*, 1638, IV, viii, 22, p. 317: "Tanta enim est Intellectus hisce in rebus pernicitas, ut definitionibus nominalibus prius perceptis, *integras & expressas simplicium dianoearum formulas*, non requirat; imo implicetur ijs potius & defatigetur, quam adjuvetur."

Scientific induction, he says, provides a means by which students can resolve questions into their simplest elements, in order to question received "nominal definitions". "If there shall be any student," he says,

> Who is either dull-witted or obstinate through having *an intellect preoccupied with opinions*, he will find this resolution of epistemonic induction into its simple notions of inestimable use.[38]

In the following chapter (*De Cautionibus in inducendo adhibendis*) Jungius notes a number of precautions to be taken when using scientific induction, and in doing so sheds extra light on how he sees it functioning. This chapter focuses on the security of different kinds of induction. The general principle is that

> Scientific induction should be done *successively and gradually*, that is, it should not ascend immediately from particulars to higher genera, but first to the lowest species, and from these to the lowest genera, and thus step-by-step to the higher.[39]

Jungius's caution, however, places strictures on this movement from particulars to lower genera, and places particular stress on the need to exhaust the realm of observable particulars and their accidental differences before deriving inductive conclusions. For example, in aphorism 2 he says:

> one should not hurry at once to the lowest species which are *absolutely adopted*, but to those *limited by accidental differences*, and one should then proceed from the limited to the less limited. *If anyone were to consider the vine in Egypt that retains its foliage through the winter, he would not immediately assert this of all vines, but of the vines which grow in Egypt or another similarly hot region.* [40]

In the following aphorism he insists that:

> a scientific induction, which infers an attribute concerning a genus [. . .] as the subject, cannot be understood to be fixed and certain until the subject parts are deduced by observation for all species.

So one cannot make the statement "wood floats in water" until one has examined the behaviour of all kinds of wood in water. In fact, he says, this conclusion can be refuted by particular instances "for *neither boxwood nor the wood of the holly-tree,*

---

[38] Jungius, *Logica*, 1638, IV, viii, 23, p. 318: "Si quis tamen auditor vel ingenio hebes, vel ob *intellectum opinionibus praeoccupatum* pertinax fuerit, usum non contemnendum habebit haec Inductionis epistemonicae in simplices suas dianoeas resolutio."

[39] Jungius, *Logica*, IV, ix, 1, p. 318: "Inductio scientifica fiat successive & gradatim, hoc est, non a singularibus statim ad superiora genera ascendatur, sed primum ad infimas species, & ab his porro ad genera infima, atque ita pedetentim ad superiora."

[40] Jungius, *Logica*, IV, ix, 2, p. 318: "Imo saepe ne ad infimas quidem *species absolutè sumptas* statim properandum est, sed ad eas *accidentalibus differentijs restrictas*, & a restrictis deinde ad minus restrictas est procedendum. Ut si quis *in Aegypto vitem pampinos per hyemem retinere animadvertat, non statim de omni vite hoc ipsum asseverabit, sed de vite, quae in Aegypto aliave similiter calida regione crescat.*"

*nor the wood of the Guaiacum nor the wood of the Brasilianum floats in water, but sinks.*"[41]

In general, he argues, "Inductions which are made in physics concerning the lowest species, are more certain than those made from genera"[42] and "Those inductions which are made about simpler things are safer and less vulnerable to [contradictory] instances, than those which concern more complex things." Thus:

> an induction by which we infer that *All salt dissolves in water, [or] all potter's clay is hardened by heat,* is more secure than one which claims *Every narrow-leaved tree sheds its leaves in winter, [or] Every horned quadruped with tallow does not grow fat with lard.* [43]

Even if a complex subject seems invulnerable to counter-instances when it is predicated, he says, it may become vulnerable to them:

> For example, *for many centuries this induction was held to be secure: a horse, an ass or a mule is an animal which can be domesticated, or suitable for human uses; and every solid-hoofed animal is either a horse, an ass or a mule. Therefore every solid-hoofed animal is able to be domesticated. But in the kingdom of the Congo discovered by the navigations of the Portuguese it has been reported that there is a solid-hoofed animal called a Zebra by the local inhabitants, which until now has not been mastered by reins, nor has been able to be tamed by any other method.* [44]

In this stress on the methodical exhaustion of particulars and their accidental differences Jungius is it variance with Zabarella, who in his *De regressu* argued that in demonstrative induction the philosopher did not need to detain himself for very long with particulars because having seen a few examples of the relevant particulars "our mind straightway notices the essential connection" (*quoniam mens nostra quibus-*

---

[41] Jungius, *Logica*, 1638, IV, ix, 3, pp. 318–319: "Inductio scientialis, quae de genere aliquo [..] ut subjecto, attributum aliquod colligit, firma ac rata haberi nequit, antequam per omnes species [. . .] subjectas partes observatione deducatur. [. . .] V. g. *Lignum in aqua natare* sive *lignum aqua levius esse, plurimum ligni specierum observatione constat, nec tamen ideo necessaria est propositio, Omne lignum in aqua natat*: Quod si quis ex *mentali terminorum examine*, quod fingunt quidam, sibi persuaserit, refutabitur *instantijs*; nam *nec Buxi, nec Aquifoliae lignum, nec Guajacinum, nec Brasilianum in aqua natat, sed subsidit.*"

[42] Jungius, *Logica*, 1638, IV, ix, 5, pp. 319–320: "5. Inductiones quae in physicis de *specie infima* fiunt, certiores sunt, quam de *genere.*"

[43] Jungius, *Logica*, 1638, IV, ix, 7, p. 320: "Inductiones item, quae fiunt in *rebus simplicioribus*, tutiores sunt, & minus obnoxiae instantijs, quam quae in *rebus magis compositis* versantur. Ut securior est Inductio, qua inferimus *Omnem salem aqua liquescere, Omnem argillam calore indurari,* quam qua colligimus *Omnem arborem tenuifoliam hyeme defrondescere, Omne cornutum quadrupes sevo non adipe pinguescere.*"

[44] Jungius, *Logica*, 1638, IV, ix, 8, p. 321: "V. g. *Iam inde a multis seculis firma fuit habita haec inductio, Et equus, & asinus, & mulus, est animal cicurabile, sive usibus humanis commodum; Atque omne solidungulum animal est vel equus, vel asinus, vel mulus. Omne igitur solidungulum animal est cicurabile. At regno Congensi Lusitanorum navigationibus detecto repertum est animal solidungulum, Zebra incolis dictum, quod tamen hactenus nec freno domari, nec alia ratione cicurari potuit.*"

*dam inspectis statim essentialem connexum animadvertit*) between them and can move quickly back to universals.[45]

While writers on method in the sixteenth century often distinguished between method as it was used to produce knowledge within a discipline (*ordo doctrinae*) and method as it was used to teach the discipline (*ordo docendi*),[46] like other Encylopaedists working in the German pedagogical milieu Jungius stresses the continuity of the procedures followed to produce knowledge and those used to teach it. For Jungius, total knowledge (*scientia totalis*) is "the system of conclusions inferred by scientific methods, together with the principles of the same methods."[47] That is, "Systematic propositions constitute the matter of 'total knowledge'." This accumulation of systematic propositions relates to didactic order (*ordo didascalicus*) as matter relates to form—that is the two orders are inextricably intertwined at the deepest level.[48]

Jungius sees Aristotle's "method" (Μέθοδος) and Galen's "didactic order" (Διδασκαλία τάξεως έχομένη) as identical concepts.[49] Just as scientific or demonstrative induction must begin with the simplest terms and gradually ascend towards more complex terms, so in didactic order:

> One should proceed therefore from *simple things* to *composite things*, and from the *less composite* to the *more composite*, unless there is some hindrance. Thus *a grammarian proceeds from letters to syllables, from syllables to words, and from these to the syntax and structure of words: likewise the Logician proceeds from notions to a proposition, first a simple one and then a complex one.* [50]

Teaching, in fact, proceeds in the same way as scientific induction: beginning with the resolution of complex questions into their simplest components:

---

[45] Jacopo Zabarella, *De regressu*, cap. 4, cit. Randall, p. 56: "Demonstrative induction can be carried on in a necessary subject matter, and in things that have an essential connection with each other. Hence it does not take all the particulars into account, since after certain of them have been examined our mind straightway notices the essential connection, and then disregarding the remaining particulars proceeds at once to bring together the universal. For it knows that it is necessary that the same relations should be embodied in the rest." (*Inductio autem demonstrativa fit in materia necessaria, et in rebus quae essentialem inter se connexorum habent. Ideo in ea non omnia sumuntur particularia, quoniam mens nostra quibusdam inspectis statim essentialem connexum animadvertit, ideoque enim necessarium esse ut ita res se habeat in reliquis.*)

[46] Reiss, pp. 201–202.

[47] Jungius, *Logica*, 1638, IV, xvi, 4, p. 348: "*Scientia totalis* systema est conclusionum per methodos scientificas illatarum, una cum earundem methodorum principiis."

[48] Jungius, *Logica*, 1638, IV, xvi, 7, p. 348: "*Materiam* itaque totalis scientiae propositiones systematicae constituunt: *Forma* est *Ordo didascalicus*."

[49] Jungius, *Logica*, 1638, IV, xvii, 2, p. 349: "Dicitur etiam *Via doctrinae* όδος διδασκαλιας, saepe *Methodus* quoque, ita ut per *methodicam institutionem* recte ordinata institutio sit intelligenda. Ipsa etiam disciplina ordine disposita, quae Galeno Διδασκαλία τάξεως έχομένη Aristoteli Μέθοδος quandoque dicitur."

[50] Jungius, *Logica*, 1638, IV, xvii, 10, p. 350: "A *simplicibus* igitur ad *composita*, & a *minus compositis*, ad *compositiora*, nisi quid obstet, progrediendum. Ita *Grammaticus a literis ad syllabas, a syllabis ad voces, ab his ad syntaxin & structuram vocum*: Ita *Logicus a notionibus ad enuntiationem, primo simplicem, deinde compositam procedit.*"

If the more composite things are better known than the simpler things then so that these simpler things might be obtained from the more composite things one should work one's way from *appearances* to *things which are hidden* (ἀπό φαινομένον ἐπι κεκρυμμενα).[51]

This outline of the application of logic to the natural sciences, from induction, through a totalised system of knowledge to "didactic order" is also the structure which Jungius applied to his scheme for a revitalised natural philosophy which he outlined in a short treatise called *Protonoeticae Philosophiae Sciagraphia* (or "Outline of a Protonoetical Philosophy"). This work, which was being circulated amongst English Baconians as early as 1640, was described in Samuel Hartlib's *Ephemerides* as "a short Idaea of the whole instauration of Philosophy".[52] There is a close relationship between this summary of Jungius's protonoetical philosophy and the fourth book of the *Logica Hamburgensis*, as can be seen from the first two aphorisms:

1. It is called Protonoetic Philosophy because it resolves every operation of the intellect into *protonoemata*, that is, irresolubile notions.
2. For it proceeds from a finite number of *protonoemata*, by means of a finite number of legitimate means, to a finite number of precisely formed notions: and in the same way from a finite number of *protophases*, or undemonstrable propositions, it infers a finite number of conclusions from a finite number of consequences.[53]

It was the protonoetical philosophy which Jungius promoted as the philosophical underpinning of his short-lived "Societas Ereneutica" in the 1620s, whose schedule of twenty-three laws (the *Leges Societatis Ereneuticae*) refers to it as the "best method" (*optima methodo*) for promoting the pursuit of new knowledge. The first law states the principal aim of the society, and in doing so describes the nature of the protonoetic philosophy:

Let it be the sole aim of our College, both to inquire into truth by reason and experience, and to reveal what is found; to free by reason and experience all arts and sciences depending on Sophistical [philosophy] and to reduce [them] to demonstrative certainty, to propagate them by correct teaching, and lastly to augment them by fertile invention.[54]

---

[51] Jungius, *Logica*, 1638, IV, xvii, 11, p. 350: "Si compositiora simplicioribus magis nota sint, ita ut haec ex illis sint indaganda, ab *apparentibus* ad *abdita*, ἀπό φαινομένον ἐπι κεκρυμμενα est enitendum."

[52] Hartlib Papers, *Ephemerides*, 1640, 30/4/56B.

[53] Jungius, *Protonoetica Philosophiae Sciagraphia*, printed in Kangro, 1968, pp. 256–270, p. 256: "1. Protonoetica Philosophia dicitur, quia omnem intellectus operationem resolvit usque in protonoemata h. e. irresolubiles notiones. 2. Progreditur enim a protonoematibus numero finitis, per modos componendi legitimos, numero finitos, ad numero finitas distincte formatas notiones: itemque a protophasibus sive propositionibus indemonstrabilibus numero finitis, per numero finitas consequentias, conclusiones infert numero finitas." On the connections between Jungius's protonoetical philosophy and book IV of the *Logica Hamburgensis*, see Meyer, 1957, p. xiii.

[54] Jungius, *Leges Societatis Ereneuticae*, printed in Martin Fogel, *Historia Vitae et Mortis Ioachimi Jungii Mathematici summi ceteraque incomparabilis Philosophi. Auctore Martino Vogelio Hamburg.* (Leipzig, 1658), p. 24: "*Scopus Collegii nostri unicus esto, veritatem è ratione & experientia tum inquirere, tum inuentam commonstrare; siue artes & scientias omnes ratione & experientia*

The laws also stipulate the teaching of the protonoetic method to new members, and in the *Sciagraphia*, Jungius clearly conceives of his philosophy as something to be taught. In aphorism 4 he says:

> Doctrine taught protonoetically compared to doctrine as it is vulgarly conceived is as speaking clearly is to stuttering or stammering, or as a codex in Plantin type is to a manuscript written with topsy-turvy, confused and tangled-up characters.[55]

The "distinction of doctrine", which is produced by "the resolution of notions", i.e. the analytic part of the inductive process, is paralleled by the "distinction of the grades of learning or understanding", which obtain in each science. These are the "empirical", the "epistemonic" and the "heuretic".[56]

At the empirical grade, the student unreflectively accepts the teachings of his master simply because they have had confirmation from experience, without giving any thought to the principles on which the precepts are based:

> for example, if [the student] learns to measure distances from afar by means of a surveying instrument [*dioptron*] and the measurement being made at close range, he discovers that the measurements made at a distance agree a number of times, he may ignore the foundations on which the rules taught by the teacher are constructed.[57]

A student who has attained the epistemonic grade, is one:

> who knows how to deduce apodictically from those things which one ought to know from a finite number of principles or *protophases* (by hypotheses, or axioms): For example, one who can demonstrate the rules for measuring distances from the sixth book of Euclid.[58]

The final, heuretic, grade is reserved for the student who:

> Has learnt the method, for solving problems which have not previously been solved and is able to discover new theorems, and to establish new rules, for example if he were able to devise new surveying instruments or new ways of measurement by a new method.[59]

---

*subnixas à Sophistica vindicare, ad demonstratiuam certitudinem reducere, dextra institutione propagare, denique felici inuentione augere.*"

[55] Kangro, 1968, p. 256: "4. Doctrina protonoetice tradita ad doctrinam vulgariter conceptam se habet ut distincte loquens ad balbutientem aut blaesum, aut ut codex Plantini typis excusus ad codicem characteribus inter se permutatis, confusis, intricatis, conscriptum."

[56] Kangro, 1968, p. 258: "6. Distinctionem doctrinae, a resolutione notionum ortam, comitatur ultro distinctio graduum discendi, cognoscendive, qui in qualibet scientia tres sunt, Empiricus, Epistemonicus, et Heureticus."

[57] Kangro, 1968, p. 258: "7. In Empirico gradu discens subsistit, qui canones, praecepta monita a praeceptore tradita ideo vera statuit, quia experientiae congrua deprehendit: v.g. si quis distantias eminus per dioptron demetiri discat et dimensionem cominus factam, eminus factae aliquoties consentire deprehendat, fundamenta tamen quibus canones a praeceptore traditi superstructi sunt, ignoret."

[58] Kangro, 1968, p. 258: "8. Epistemonici gradus particeps est, qui ea quae scire debet ex principijs sive prophasibus (hypothesibus, axiomatibus) numero finitis, apodictice deducere novit v.g. qui distantias eminus metiendi canones e libr. 6 Euclidis demonstrare potest."

[59] Kangro, 1968, p. 258: "9. Heureticum gradum obtinet, qui methodum novit, qua problemata, ante non soluta solvere, nova theoremata invenire, novos canones condere queat, v.g. si quis nova dioptrica instrumenta vel novos emetiendi modos nova methodo excogitare valeat."

It is interesting to note that Jungius takes his examples from applied mathematics, which marries both sides of his scientific edifice: the logical structures of demonstration and their application to the realm of actual experience, that is to say, induction by means of empirical theorems. The model of mathematical demonstration is foregrounded much more in the *Sciagraphia*, and it was clearly a vital component in his concept of protonoetic philosophy. This is particularly striking in aphorism 13 where Jungius criticises the peripatetics for beginning the study of philosophy with logic, which, he says, is "the occasion of many evils". In doing this, he argues:

> They put reflective understanding in advance of direct reasoning, and inculcate the Heuretic grade before the Epistemonic grade. Physics, rightly ordered, no more requires a preparatory course in Logic than Euclidean geometry.[60]

While this might seem curious coming from a professor who had written an introductory logic for use in the schools of Hamburg, his antagonism concerns what he perceives as the defects of the Aristotelian logic then being taught:

> The peripatetics do not say a word about the resolution of notions, the most useful and greatly necessary part of Logic, and they teach hardly anything good in Logic beyond the first and second Analytics, but these things themselves are by no means necessary for the acquisition of the sciences, [and] they are far from adequate for the refutation of opinions.[61]

As a student at Padua, and an avid (if critical) reader of Zabarella, Jungius must have known that this is not the case with Zabarella's demonstrative regress, which combines resolution and composition of notions in precisely the way in which Jungius suggests. The anti-Aristotelian tenor, in a thinker who so clearly works within the sphere of the Aristotelian conception of *scientia*, is quite surprising, and becomes more trenchant as the treatise progresses. "Those who are eminent in peripatetic philosophy", he says,

> delight in that Exodic kind of teaching, that is they proceed by disputing on both sides [of a question], doubting, solving, opposing, replying, refuting, distinguishing [and] reconciling, virtually nowhere do they propound his dogmas without ambiguity, nor do they prove them with clear apodictical proofs.[62]

---

[60] Kangro, 1968, p. 260: "13. Iidem pessimo consilio Philosophiae studia a Logicis auspicantur, quod multorum malorum occassio est. Reflexivam enim cognitionem hac ratione praemittunt directae, et gradum Heureticum inculcant ante Epistemonicum. Physica recte concinnata non magis quam Geometria Euclidaea, praecursu indiget Logices."

[61] "12. Peripatetici notionum resolutionem partem Logicae tum utilissimam, tum maxime necessariam, ne verbo quidem attingunt, neque boni fere quicquam in Logicis docent praeter pauca e prioribus Analyticis, verum et haec ipsa ad scientias adquirendas minime necessaria sunt, ad opiniones vero refutandas longe non sufficiunt."

[62] Kangro, 1968, p. 260: "14. De Exodico docendi genere gaudent, qui in Peripatetico [*Kangro*=Peripato] eminent, hoc est disputando in utramque parte, dubitando, solvendo, opponendo, replicando, refutando, distinguendo, conciliando procedunt nusquam fere dogmata sua absque ambiguitate proponunt, nusquam per liquidas apodixes comprobant."

Jungius is clear about the intended audience of his polemic. It is for those like himself, who cannot contain their natural philosophy within the boundaries defined by the schools, but wish to investigate nature experimentally:

> With errors so frequent in vulgar Philosophy, apodictically uncovered, honest Philosophers can no longer bear to wander in the labyrinth of the peripatetics with the idle, but prefer modest and plain solid knowledge to prolix and splendid opinions, and they eagerly and constantly bestow money and effort on observations so that they may increase and extend it.[63]

The aim of both the *Sciagraphia*, and the fourth book of the *Logica Hamburgenis* was to show how it might be possible to reduce all the arts and sciences to demonstrative certainty.[64] Beginning with an essentially Aristotelian conception of *scientia*, and retaining the Aristotelian logical superstructure of his scientific induction, Jungius brought to it his experience as a student of medicine, mathematician, natural historian and chemical experimenter, and sought to extend it to become the basis for what he saw as a better and more effective philosophy of nature.

As Timothy Reiss has suggested, René Descartes may well have gone through a similar journey, taking as his starting point the methodological debates of the sixteenth century, viewed, "through a Toletan or [. . .] Zabarellan filter" before "making something different (and more complicated) out of them".[65] Jungius himself saw profound similarities, but also profound differences between himself and Descartes. In a long letter to Descartes's patron princess Elizabeth in March 1655, he made an unflattering comparison between the *Logica Hamburgensis* and Descartes's *Principia Philosophiae* (which had been published 6 years after the *Logica* in 1644).[66] Jungius clearly sees parallels between what Descartes was trying to achieve in that work and his own protonoetic philosophy. He criticises Descartes, however, for having confused logic and metaphysics: "I would have preferred to have discussed physical matters with you, from book two of the *Principia*," Jungius writes,

> (those parts which I call the four books in my writings). But as matters stand it is requisite that first we should concentrate on logical principles, otherwise the trouble taken over the rest will be in vain. I strongly disapprove of Descartes – a man who is otherwise eminent – because of the little care that he has taken over logic, which he almost confuses with metaphysics, and reflective knowledge with correct knowledge. The first book of the *Principia* is totally concerned with metaphysics, combined with many things pertinent to logic. Article 45 propounds definitions of clear and distinct perceptions, which if you want to see for yourself whether it has been well handled, you may compare the definition which has been given of a distinct notion in book I, final chapter, aphorism 4 of the *Logica Hamburgensis*.[67]

---

[63] Kangro, 1968, p. 268: "29. Erroribus tam densis in Philosophia vulgari apodicticè detectis Philosophi ingenui in otiosis Peripateticorum labyrinthe obambulare amplius non sustinebunt, sed prolixis et splendidis opinionibus modicam et tenuem, solidam tamen scientiam praeferent, eamque ut exaugeant et extendant sumptus et operas in observationes alacriter et constanter conferent."

[64] See fn. 54 above.

[65] Reiss, p. 211.

[66] Joachim Jungius to Elisabeth Pfalzgräfin bei Rhein, 23 March 1655, printed in Risse, *Additamenta*, pp. 217–219.

[67] Risse, *Additamenta*, p. 217: "Mallem sane tecum conferre de physicis rebus ex libro II. de Princip[iis] (libros IV in schedis meis voco, quos illae partes). Verum oportet prius nos in logicis

He also finds it "intolerable" that Descartes should reduce affirmation and negation to modes of volition, and he advises her to consult instead his discussion of the three basic kinds of mental operation on which his own logic is founded.[68] With all the curmudgeonly reluctance one might expect from an older academic who balefully appraises the work of a younger academic star, he concedes that Descartes *Est iuvenis ingeniosus, sed nimis festinans* ("is an ingenious young man, but a little overhasty")! Although this judgement seems implausibly comic now, in Jungius's own day his own intellectual stock was high: the English Baconians saw Jungius, Bacon and Descartes as intellectual equals, and Leibniz (in a letter to Christian Philipp in December 1679), claimed that "There are several errors in his [i.e. Descartes's] metaphysics, and he has not understood the veritable source of truths nor that general analysis of notions which, in my view, Jungius understood better than him".[69] He was certainly well thought-of in English scientific circles, and in 1668 Henry Oldenburg, writing to Jungius's editor Martin Fogel, praised Jungius's *Phoranomica*, as a "great store of observations and [...] considerable treasury of mechanical science." The Royal Society, Oldenburg told Fogel, was striving for the promotion of "the solid method of philosophizing of Bacon and Jungius."(*Verulamii et Jungii solidam philosophandi methodum*).[70]

Whatever else one might want to say about the respective merits of the Cartesian and the Jungian methods, this much is clear—that both men sought to institute a method by which clear and exact knowledge could be obtained, and in doing so both felt the anxiety of scholastic influence. In a posthumous collection of Jungius's "doxoscopies" (or critical reviews of philosophical doctrines), the *Doxoscopiae Physicae minores* published in 1662, the editor, Fogel, passes on an autograph note (*scheda*) which he claims to have found stuck to the front of the original manuscript.

principiis convenire, alioquin frustra sumetur labor in reliquis. Valde enim hoc improbo in Cartesio, viro alias magno, quod adeo exiguam logicae curam habet, eamque quasi cum metaphysica confundit, scientiam reflexivam cum recta. Liber primus de principiis totus occupatur in metaphysicis, admixtis nonnullis ad logicam pertinentibus. [Lib. I] Articulo 45 proponit perceptionis clarae et distinctae definitiones, quae utrum bene sit traditae, in tuo intellectu experiri poteris, praesertim si conferas definitionem notionis distinctae in Hamburgensi Logica lib. I, c. ult. § 4. traditam." (see Jungius, *Logica*, 1638, I, xv, 4, p. 101: "*Notio distincta* est cum rem definitione vel accurata descriptione explicatam concipimus, sive cum rei notionem ita formamus, ut, quomodo formata ac composita sit, non ignoremus. Ut *si Circulum definiam superficiem planam peripheria terminatam, et Peripheriam, lineam recurrentem cujus quodvis punctum ab uno aliquo intra eam puncto, quod Centrum dicatur, aequaliter distet*; *Distantiam* denique *puncti à puncto, lineam rectam punctis interjectam*.").

68 Risse, *Additamenta*, p. 219: "Id vero intolerabile est, quod affirmationem et negationem vult esse modos volendi, i.e. species volitionis. Art. 32. Haec omnia eo, ne patiaris te ab Hamburgensi logica abduci, quae tribus mentis operationibus superstructa est."

69 G. W. Leibniz to Christian Philipp, December 1679, cit. Risse, *Additamenta*, p. 381: "Il y a plusieurs erreurs dans sa [Descartes] metaphysique, et il n'a pas connu la veritable source des verités ny cette analyse générale des notions que Iungius à mon avis a mieux entendu que luy."

70 Henry Oldenburg to Martin Fogel, 4 April 1668, *The Correspondence of Henry Oldenburg*, ed. by A. Rupert Hall and Marie Boas Hall, 13 vols (Madison: University of Wisconsin Press, 1965–1986), IV, pp. 302–303.

In it Jungius presents a clear example of what must have been the emulatory impulse of many of the new philosophers, reared on the Aristotelianism of the schools:

> Aristotle never put forward his doctrines before he had refuted contrary positions. I will use the same method. I will put forward my judgement once I have destroyed those of Averroes and Zabarella.[71]

**Acknowledgements** I would like to thank Dr Peter J. Forshaw, British Academy Postdoctoral Fellow at Birkbeck, University of London, who read an earlier draft of this essay and made many helpful suggestions.

[71] *Joachimi Jungii Lubecensis Doxoscopiae Physicae minores, sive Isagoge Physica Doxoscopica. In qua praecipue Opiniones in Physica passim receptae breviter quidem, sed accuratissime examinantur. Ex recensione et distinctione M[artinii] F[ogelii] H[amburgensis] cujus Annotationes quaedam accedunt* (Hamburg, 1662), "Annotatio generalis" sig. α2v: "*Nec contra Auctoris mentem* praemisero Doxoscopicas Systematicis ejus Scriptis, cum ipse hoc Ordine: se usurum promittat in Scheda, quae adhaesit Praelectionum Autographo: *Aristoteles nunquam proponit suam Sententiam, antequam refutavit Contrarias. Ego utar eadem Methodo. Propositurus sum meam Sententiam, ubi Averroem & Zabarellam destruxero.*"

# *Scientia* and the Sciences in Descartes

**Tom Sorell**

Descartes appears to have more than one conception of science. First, there is the relatively exacting conception that he associates with *scientia*. This is the conception that comes into play in a well-known passage from the Replies in which he explains how an atheist can and can't know that the angles of a triangle add up to two right angles: the atheist cannot know in the sense of having an unshakeable conviction that the two angles add up to two right angles, but the atheist *can* know—grasp for as long as no searching doubt is conjured up—the truth that the two angles add up to two right angles. So, exactingly conceived, science is the ultra-stable grasp of truth, ultra-stable because supported by general reasons for confidence in the methodically applied human intellect. Less exactingly conceived, science is successful problem-solving or explanation in terms of a small number of widely applicable "simple" notions—shape, size, number position or motion in the case of physics. Solutions or explanations along these lines do not have to command unshakeable conviction. For one thing, they can often be rivalled by other would-be explanations that cohere well with a prevailing set of received ideas. Nevertheless, if they answer a question or make intelligible some data in a systematic and economical way, meeting some acceptable standards of clarity, a solution or an explanation can have considerable authority. Science along these lines is illustrated in the three *Essais* introduced by the *Discours*, and in *Le Monde*.

What is the relation between specimens of the exacting and less exacting conceptions of science? Are *scientia*-ranking items of knowledge more truly science than what one finds in the *Essais* and *Le Monde*, or do *scientia*-ranking items of knowledge include what one finds in the *Essais* and *Le Monde*? I shall suggest that when understood in the light of the *scientia* that one gets from Descartes's metaphysics, the otherwise merely morally certain findings of the *Essais* and *Le Monde* are raised to the level of *scientia* themselves. But this raises the question of whether *scientia* in the form of metaphysics is a higher form of *scientia* than *scientia* in the form of e.g. physics. The answer to this question to be found in Descartes's writings is more complicated than one might expect. On the one hand, metaphysics is the most

T. Sorell (✉)
University of Birmingham, Birmingham, UK

T. Sorell et al. (eds.), *Scientia in Early Modern Philosophy*, Studies in History and Philosophy of Science 24, DOI 10.1007/978-90-481-3077-1_5,

basic of the sciences, and so might be thought to be a more privileged or higher form of *scientia* than *scientia* in the form of physics, medicine, mechanics and so on. On the other hand, the more derivative natural sciences suit our nature better, and are better objects of everyday human attention than metaphysics, while morals is the highest science in the sense of having the most direct use in our perfecting ourselves. Again, all of the sciences have something in common—the production of synoptic understanding in an order that starts with the "simple", so that metaphysics is not *sui generis*.

## I *Scientia* and Knowledge

The *Meditations* provides a proof, in the face of scepticism about all of learning—that *scientia* is possible for human beings (see esp. AT VII 69, 71; CSM II 48, 49). *Scientia* is a special kind of knowledge. It is systematic knowledge of a lot of related truths. So it contrasts with intuitions of single truths. Intuitions of single truths are knowledge; and so long as the attention is fixed, intuitions do not depend for their certainty on knowledge of God; but the attention naturally wanders, and the certainty of any one truth can be shaken by the principles for calling into question whole classes of belief that Descartes introduces in Meditation One (AT VII 18; CSM II 12). So intuition needs to be backed up by something that can overturn those principles: namely the certainty that God exists and is no deceiver. *Scientia* is knowledge of a system of truths that incorporates this assurance against general uncertainty. And crucially, once it is arrived at, it is not open to reconsideration through doubt, not even through the doubt that our minds might be defective.[1] *Scientia* is an unshakeable vision of a system of truth because it is a vision that includes *scientia* that God exists and is no deceiver. Being unshakeable and true, it resembles divine knowledge itself. God could not know now and later lose his knowledge, for he wills all that is eternally true all at once and for all time, and cannot fail to know what is therefore willed to be true for all time. In other words, God's knowledge is unshakeable. Of course, human *scientia* and divine *scientia* differ much more than they resemble one another. To the extent *we* have *scientia,* it is by disciplining the attention, ordering its objects from simple to complex and getting used to seeing totalities in the light of the simple. God's purchase on the objects of *scientia* is entirely different. For one thing, those objects are his creations, and he knows the natures of things by deciding what they will be, not by fixing his attention on them.

If God's immutability is both a model for human *scientia* and something we have to be aware of in order to acquire *scientia*, cannot Descartes claim with justice to have made knowledge of God essential to the knowledge of nature, just as the

[1] For more on the distinction between *scientia* and intuition (crucial to the issue over the Cartesian Circle) see A Kenny, *Descartes: A Study of his Philosophy* (New York: Random House, 1968), ch.8, and J Tlumak, "Certainty and Cartesian Method" in M Hooker, ed. *Descartes: Critical and Interpretive Essays* (Baltimore: Johns Hopkins University Press, 1978) pp. 40–73.

theologians would have wanted him to do? The claim has some, but only some, plausibility in the context of the *Meditations*, as we shall see. But it is much weaker in connection with the *Discours* and *Essais* and even weaker in *Le Monde*.

Let us begin with the *Discours*. Not only does Descartes pass very quickly over metaphysical truths in Part Four, allowing the reader to infer from the brevity of his treatment that they do not matter much, but he also makes the very striking claim that the evident adequacy of the explanations of things in *la Dioptrique* and the *Meteores* is on its *own* a proof of the principles of Cartesian natural science put forward in at least the two first *Essais*. This is how we might understand the following famous passage from the end of the *Discours*.

> Should anyone be shocked at first by some of the statements I make at the beginning of the *Optics* and *Meteorology* because I call them 'suppositions' and do not seem to care about proving them, let him have the patience to read the whole book attentively, and I trust that he will be satisfied. For I take my reasonings to be so closely interconnected that just as the last are proved by the first, so the first are proved by the last, which are their effects. It must not be supposed that I am here committing the fallacy that the logicians call "arguing in a circle". For as experience makes most of these effects quite certain, the causes from which I deduce them serve not so much to prove them as to explain them; indeed, quite to the contrary, it is the causes which are proved by the effects (AT VI 76; CSM I 150).

In the same vein, there is this passage from the letter to Mersenne of 27 May 1638:

> You ask if I regard what I have written about refraction as a demonstration. I think it is, in so far as one can be given in a field without a previous demonstration of the principles of physics by metaphysics—which is something I hope to do some day but which has not yet been done–and so far as it has ever been possible to demonstrate the solution to any problem of mechanics, or optics, or astronomy, or anything else which is not pure geometry or arithmetic. But to require me to give geometrical demonstrations on a topic that depends on physics is to ask me to do the impossible. And if you will not call anything demonstrations except geometers' proofs, then you must say that Archimedes never demonstrated anything in mechanics, or Vitellio in optics, or Ptolemy in astronomy. But of course nobody says this. In such matters people are satisfied if the authors' assumptions are not obviously contrary to experience and if their discussion is coherent and free from logical error, even though their assumptions may not be strictly true. I could demonstrate, for instance, that even the definition of the centre of gravity given by Archimedes is false, and that there is no such centre; and the other assumptions he makes elsewhere are not strictly true either. . . .but that is not a sufficient reason for rejecting the demonstrations. . . .(AT II 141–142; CSM III 103).

Demonstrations can be demonstrative, in other words, if they meet conditions far less exacting than those for metaphysical certainty. Or, in other words, there can be science without *scientia*.

Now it might be thought that whatever concessions are made to science lacking in metaphysical certainty in the *Discours* and *Essais*, these concessions would be withdrawn in the *Meditations*. But this is not quite what one finds. If anyone lacks metaphysical certainty, according to the principles in the *Meditations*, it is someone who is ignorant of God's existence, or someone who denies it: an atheist. But does the atheist necessarily lack knowledge or science, science in the sense of knowledge of something that results from demonstration? More than one set of objectors thought this was implausible (cf. AT VII 124; CSM II 89; AT VII 414; CSM II 279), and the

Replies do not entirely disagree. Just as there can be a genuine demonstration that involves some incidental falsehood and that lacks metaphysical grounding, so there can be a kind of grasp of truth available even to those who think there is no God.

Descartes never even denied that in some sense of "knowledge" the atheist could have knowledge; what he denied was that the atheist could have the preferred kind of knowledge called *scientia*:

> The fact that an atheist can be 'clearly aware that the three angles of a triangle are equal to two right angles' is something I do not dispute. But I maintain that this awareness of his is not true knowledge, since no act of awareness that can be rendered doubtful seems fit to be called knowledge (AT VII 141; CSM II 101).

> As for the kind of knowledge possessed by the atheist, it is easy to demonstrate that it is not immutable and certain... (AT VII 428; CSM II 289).

As if knowledge that was not immutable or incapable of being reconsidered critically could not *really* be knowledge. All we need interpret Descartes as saying is that the atheist is bound to lack *scientia*. We do not need to interpret him as saying that *scientia* is the only knowledge there is. There is also a lesser kind of knowledge.

This category of non-ideal knowledge has an interesting range of application. Apart from the atheist, it extends to the scientifically open-minded believer who does not know Descartes's metaphysics, including churchmen who were instructed in optics and meteorology by the *Essais* and would have been instructed by *Le Monde* in physics. More important, it applies to Descartes himself before he discovered his proofs of metaphysical truths in 1630. The fact that Descartes himself had only imperfect knowledge before 1630 may show that the scientific discoveries he made before 1630 were not *scientia* in the sense of the *Meditations.* But surely they counted as science. The fact that these discoveries might have been rendered doubtful or disputed by critics does not seem to make them count any less as science, just as the fact that the demonstrations of Archimedes might be corrected does not show that they were not demonstrations. On the contrary it is the honorific sense of "science"—science in the sense of *scientia*—and the honorific sense of "demonstration"—"demonstration" in the sense of geometrical demonstration—that seems unduly narrow or stipulative—even, at times, from Descartes's own point of view as scientist. Up to a point this narrowness is the direct result of the grounding of Descartes's science in the doctrine of God's nature.[2]

The point can be made in a different way. Before 1630—before science was metaphysically grounded—Descartes may be said to have developed a merely provisional philosophy of science broadly analogous to the *morale par provision* of the

[2] It is true that in the passages just quoted, as at the very end of Meditation 5, Descartes is under the influence of the Platonic idea that while true beliefs can come and go, knowledge has got to be stable and fixed, tied down by the reasons for it. The idea is given a distinctively Cartesian gloss when stability is understood in terms of immutability of mind, and where the ideal mind is the immutable mind par excellence—God's. To the extent that Descartes's scientific practice requires a less metaphysically grounded philosophy of science than he constructed to please the theologians—to that extent—and I would say it is a considerable extent, Descartes can be said to have failed to make Cartesianism a better ally of Catholicism than Thomism.

*Discours,* The provisional philosophy of science is permissive about what to include as science. It includes the imperfect demonstrations of Archimides and Ptolemy. The official or strict philosophy of science is the theologically correct one in which *scientia* is the human counterpart of divine knowledge. It excludes discoveries that can be rendered doubtful or that contain elements of falsehood and probably only contains the Cartesian sciences. The more restrictive philosophy of science is not necessarily a better or even more authentically Cartesian philosophy of science. Unshakeable certainty in physics and mathematics is intelligible as an ideal, of course, especially if divine *scientia* sets the pattern for human cognitive perfection. But openness to the opinion of others at least as intent upon the truth as oneself is also consistent with a programme of enlarging knowledge. In the *Discours* Descartes seems to have been keen to display such open-mindedness, inviting objections from his readers and promising to respond to them. The open-mindedness seems to have been more than a pose, as Descartes does seem to have seen the value of the work of his contemporaries—Harvey, for example, or Galileo, even Hobbes when it came to civil science. The emphasis on unshakeable certainty, privately arrived at, seems to have started with the composition of the *Meditations.*

## II Two Conceptions of Science Reconciled

Descartes goes some way toward reconciling the restrictive with the non-restrictive conceptions of science even in the *Discourse* and *Essais.* Part Six of the *Discours* reports how far he was able to get in physics without having to make the particular observations necessary to solve the most specific problems:

> First I tried to discover in general the principles or first causes of everything that exists or can exist in the world. To this end I considered nothing but God alone, who created the world; and I derived these principles from certain seeds of truth which are naturally in our souls. Next I examined the first and most ordinary effects deducible from these causes. In this way it seems to me, I discovered the heavens, the stars, and an earth; and, on the earth, water, air, fire, minerals, and other such things, which, being the most common of all and the simplest, are consequently the easiest to know. Then, when I sought to descend to more particular things, . . .I thought that the only way of making these bodies useful to us was to progress to the causes by way of the effects and to make use of many special observations (AT VI 64: CSM I 144).

By contrast with *Le Monde*, the *Essais*, at least *La Doptrique* and *Les Meteores*, plunge directly into the explanation of particular things leaving out the more general principles of physics, but hinting at them. Although the explanations are self-contained and real demonstrations, as Part Six of the *Discourse* and the letter to Mersenne about refraction make plain, they leave between the lines the fundamental principles of physics and the background principles about the nature of God. This is the ghost context for the demonstrations, the principles whose absence renders some elements of the explanations mere "suppositions". Had they been inserted, the principles would have turned the demonstrations in *La Dioptrique* into *scientia.* In a sense Descartes wants to have it both ways. He wants to say that the explanations in

*La Dioptrique* are true demonstrations and free-standing, the "suppositions" being proved by their power of comprehending the data to be investigated. He also wants to say that the suppositions lack support that the principles of the suppressed physics would have supplied.[3] In a sense he *succeeds* in having it both ways, because the suppressed *Le Monde* is present in the *Discours*, bringing the support of its unstated first principles to bear on the explanations of things like refraction and the rainbow.

Because of the odd literary devices in the *Discours*, and the existence of the suppressed *Le Monde*, the specimens of physics in the *Essais* are much more like *scientia*, I am suggesting, than at first appears. The *Discours,* at least as I am interpreting it, reconciles the restrictive and non-restrictive conceptions of science by implying that specimens of the latter are really only edited extracts from the former. As edited extracts, they are intelligible on their own, but lack starting points in the simple that would make them ideally or fully intelligible. On this reading there is only one kind of physical science—*scientia*—a deduction starting from the simple nature of God, but the explanation of particular phenomena can make a sort of sense without its entire context in *scientia* being evident.

Now it is true that in *Le Monde* Descartes is not yet working with the concept of *scientia* to be found in the *Meditations*. The idea of God's nature in play in the suppressed physics treatise, especially in the crucial sixth and seventh chapters, is not something arrived at by taking doubt to its limits. Nor is God's non-deceivingness particularly insisted upon. When Descartes proposes laws of nature for an imaginary universe in the light of God's nature, it is God's immutability that is crucial (cf. AT XI 38; CSM I 93), not his benevolence, or, in particular, his non-deceivingness. Only three laws of nature have to be assumed to operate in this imaginary universe (AT XI 47; CSM I 97). Further rules, which "comprehend in a concise way all of the effects of nature", follow from these three and the eternal truths of mathematics, according to Descartes. As for the eternal truths, God himself has taught us these (ibid.). Descartes might have said that God being no deceiver, the eternal mathematical truths cannot be false if God teaches them to us. But what he actually says is that

> The knowledge of these truths is so natural to our souls that we cannot but judge them infallible when we conceive them distinctly, nor doubt that if God had created many worlds, they would be as true in each of them as in this one (ibid,).

This may actually contradict the distinctive Cartesian metaphysical thesis that the eternal truths are subject to God's will. In any case, we are not given a reason in *Le Monde* for thinking that the naturalness of the eternal truths to our minds is a reason for having confidence in them: in the passage above, Descartes simply *asserts* that we have knowledge of them, certainly not meeting the standard of certainty

---

[3]And I have called them "suppositions" simply to make it known that I think I can deduce them from the primary truths [first deduced from considerations about God alone] I have expounded above [i.e. at AT VI 64]; but I have deliberately avoided carrying out these deductions in order to prevent certain ingenious persons from taking the opportunity to construct, on what they believe to be my principles, some extravagant philosophy for which I will be blamed (ATVI 76; CSM I 150).

erected by the *Meditations*. Nor is anything done to establish beyond doubt that, God being immutable, the laws of nature are stable. *Le Monde* does acknowledge that physics depends on metaphysics in the sense that the laws of nature depend on God's immutability, and God's immutability belongs to metaphysics (AT XI 38; CSM I 93), but metaphysics does not at this stage make for unshakeable conviction. Instead of being a matter of something that resists all doubt, even metaphysical doubt, God's immutability is simply common knowledge, something that "everyone must know" (ibid).

It is not until the *Principles of Philosophy* that the physics clearly bears the marks of *scientia* in the sense of Descartes's later writings. The *Principles* drops *Le Monde*'s device of the imaginary universe, but it keeps the laws of nature and makes explicit the further rules alluded to in Chapter 7 of *Le Monde*. When it comes to the tricky question of how other things that might be deduced from the laws and further rules—including explanations of particular terrestrial phenomena—can be certain, he changes the answer given in *Le Monde*. It is no longer that the God-given eternal truths plus some uncontroversial metaphysics about God guarantee the consequences. Rather, it is the following consideration:

> We would seem to be doing God an injustice if we suspected that causal explanations deduced in this way [i.e. deduced from evident principles by mathematical reasoning] were false. For this would imply that God had endowed us with such an imperfect nature that even the proper use of our powers of reasoning allowed us to go wrong ( Pt III, §43; AT VIII A; CSM I 99).

Here we have the mature metaphysics being called upon to do nothing less than underwrite the specific applications of Descartes physics.[4] The metaphysics even more clearly underwrites the general parts of the physics, since Part Two of the *Principles*, where the more general parts of the physics are supposed to begin,[5] incorporates quite a lot of the material of Meditation Six.

It looks as if the body of the *Principles* bears out what is said in two passages in the 1647 Preface to the French edition. First,

> . . .I took the being or existence of this thought as my first principle, and from it I deduced very clearly the following principles. There is a God who is the author of everything there is in the world; further, since he is the source of all truth, he certainly did not create in us an understanding of the kind which would make us capable of making a mistake in its judgements concerning the things of which it possesses a very clear and distinct perception. These are all the principles I make use of with regard to immaterial or metaphysical things, and from them I deduce very clearly the principles of corporeal things, namely that there are bodies which are extended in length, depth and breadth, and which have various shapes

[4]Now it may also be true that the *Principles* preserves a version of what is claimed in the *Discours*—that for effects to be comprehended by principles, especially unexpectedly, is for the principles to be proved. But for someone to see only that principles are proved in this way is not to have *scientia* about the effects. I will return to this point.

[5]"I divided the book into four parts. The first contains . . .what may be called "first philosophy" or "metaphysics" . . .. The other three parts contain all that is most general in physics". Preface to the French edition of the *Principles* (1647) (AT IXB 16; CSM I 187).

and move in various ways. Here, in total are all the principles which I use to deduce the principles of other things (AT XI B; CSM 184).

The second passage states the famous metaphor of the tree of philosophy:

The first part of philosophy is metaphysics, which contains the principles of knowledge, including the explanation of the principal attributes of God, the non-material nature of our souls and all the clear and distinct notions which are in us. The second part is physics, where, after discovering the true principles of material things, we examine the general composition of the entire universe, and then, in particular, the nature of this earth and all the bodies which are most commonly found upon it, such as air, water, magnetic ore and other minerals.

He goes on to describe the more derivative parts of philosophy, comparing philosophy as a whole to

A tree. The roots are metaphysics, the trunk is physics, and the branches emerging from the trunk are all the other sciences.

Our concern is with metaphysics and physics. The second *is*, it seems, rooted in the first. That is, the certainty of the principles of material things and their general applicability even to particular phenomena, are vouched for by Descartes's metaphysics. It is because the certainty of metaphysics rubs off on the physics, and the certainty of metaphysics and physics rubs off on the more derivative sciences, that the whole of philosophy can be stable enough to count as *scientia*.

Doesn't this way of thinking make metaphysics, whose certainty is not derivative, and whose certainty contributes to the certainty of other sciences, a kind of science *par excellence*? Isn't the metaphysics more truly *scientia* than any other branch of science?

Yes and No. On the one hand, there can be no doubt that metaphysics is fundamental in the scheme of the sciences; on the other hand, metaphysics follows the same "logic", in Descartes's sense of "logic", as the more derivative sciences (cf. e.g. AT IX B 14; CSM I 186), and there is a sense in which this "logic" rather than metaphysics might be the key to whether something counts as *scientia*. Again, there is a sense in which metaphysics is less wholesome a pursuit for human beings than the more derivative sciences, however good it may be once in a lifetime for an intellect searching after truth. In this sense, too, metaphysics lacks pride of place among the sciences. "Morals" is the more likely claimant to the title of "most beneficial science" and its being the most derivative science is not a ground for thinking that Descartes regards it as second- or third- rate. On the contrary, in the Preface to the 1647 edition of the Principles, he calls morals "the ultimate level of wisdom" (AT IX B 14; CSM I 186).

## III Logic and a Logic of Science

"Logic" is not given a place in Descartes's tree of philosophy, but this may be because it lacks a subject matter of its own. Metaphysics is the science of immaterial things, including God and finite minds; physics is the science of material things, and medicine and morals are sciences of human beings as unions of minds and bodies. Logic, on the other hand, is the general method for moving between the more and less intelligible, or between the simple and the complex, in any subject matter whatsoever. Such a logic was traditionally displayed by geometry, but it was not only appropriate there. As Descartes says in the *Discours,*

> Those long chains composed of very simple and easy reasonings, which geometers customarily use to arrive at their most difficult demonstrations, had given me occasion to suppose that all the things which can fall under human knowledge are interconnected in the same way. And I thought that, provided we refrain from accepting anything as true which is not, and always keep to the order required for deducing one thing from another,there can be nothing too remote to be reached in the end or too well hidden to be discovered (CS 120; AT VI 19).

This passage has led some readers of the *Discourse* to think that, for Descartes, natural scientific explanations are supposed to be laid out as formal geometers' proofs, and that they are supposed to reach conclusions as rigorously as in Euclid, and on the model of theorems being derived from axioms. We have already seen Descartes denying this. Far less than a mathematical demonstration suffices for demonstration in natural philosophy, according to him.

What geometry and the other natural sciences have in common is the order of moving from the maximally intelligible to the less intelligible by steps that are intelligible and clear within the limits permitted by the subject matter under discussion. In the case of Descartes's own scientific essays, demonstration does not take the form of pages and pages of proof, but, typically, the elaboration of mechanical models of natural phenomena under investigation. Visible differences are like the differences in shape of tangible objects detected by a blind man's stick (CSM I 153; AT VI 85); the action of light on an illuminated object is like the action of wine in a full wine vat that can move toward either of two holes in the vat (CSM I 155; AT VI 87–88), and so on. If the models are oversimplified or inept in some other way, the explanation will fail. If, on the other hand, the model seems to capture the phenomenon under discussion, then the way the model works will illuminate (by indicating a possible way of working of) the phenomenon, too. The relation of the model and its workings to the phenomenon under explanation is not "deductive" in any familiar, 21st century, sense, but it is "deductive" in *Descartes's* sense: a lot of different considerations are ordered from highly intelligible to less intelligible in a way that can be taken in by a continuous mental sweep free of doubt or unclarity. At the end of the deduction, the less intelligible thing—e.g. the action of light—is supposed to make sense against the background of the simpler considerations—including the way a mechanical model works.

Far from depending on long chains of reasoning or mathematical formulae, Cartesian explanations are often strikingly informal. If they convince, it is often through the perception of similarity between a mechanical model and a selected phenomenon, rather than through knock down proof, or through the successful prediction of precisely measured effects from certain initial conditions.[6] Again, a Cartesian explanation purports only to find a possible cause of phenomena assumed to be producible in more than way. So a Cartesian explanation does not claim that the phenomenon in question *must* have been produced in the way it describes, or that it could only have been produced that way. There are lots of ways that an omnipotent God could have produced given effects (CSM I 144; AT VI 64–65). Descartes's method does require explanations or explanatory models to be at least as complex (to comprehend as many considerations) as the phenomenon under investigation or the problem to be solved. He does demand that the complex be understood in the light of the "simple" or fundamental. And what is simple or fundamental relative to a problem or phenomenon in natural science is usually something to do with size, shape, position, path or speed. But these conditions can be met and an explanation fall far short of being deductive in the logic-book sense or even conspicuously mathematical. When Cartesian explanations in natural science are convincing, then, that is not because they are compelling in the same way as geometrical proofs, or because they are special cases of mathematical proof.

The simples of natural scientific explanation were not the ultimate simples, and there were things—revealed by what is called "philosophy" in the *Discours*, or "first philosophy" or "metaphysics" in later writings—that were simpler still and capable of illuminating the elements or simples of the natural sciences. The ultimate "simple" is nothing other than God's nature, and we hit upon this by the methodological doubt that leads to the "cogito" and to reflection on the mind's ideas. There is more than one reason why Descartes brings in the ultimate simples. He wants to appear theologically orthodox, and to embed his version of mechanical science in a metaphysics that supports theological orthodoxy. He wants to present the science he is proposing as systematic, rigorous and general. He wants the explanatory shortcomings of the Aristotelian alternative to his own science to be connected with its reliance on the senses, and he wants to remind the theologian supporters of this Aristotelian science of its heathen origins. He wants to point out that the "simples" of natural science are just as sense-independent as the simples of metaphysics. Finally, and contrary to Aristotelianism, he wants to show that the simples of metaphysics belong to the same series as the simples of the natural sciences when the problem submitted to Descartes's method is the overarching one of how the whole range of scientific problems can be successfully solved by the human mind. Descartes brings in the most basic simples partly to show how science can be successful after a long run of failure in the hands of the Aristotelians. It can be successful—comprehend many observations with a small number of principles—when it is geared to the assumption that the most basic properties of the world are mechanical ones, and

[6]For extended discussions of the difference between Descartes's official method and the scientific practice that is supposed to apply it, see Doren Recker, "Mathematical Demonstration and Deduction in Descartes's Early Methodological and Scientific Writings", reprinted in T. Sorell, ed. *Descartes* (Aldershot: Ashgate, 1999), pp. 3–24, and, in the same volume, William R. Shea, "Descartes: Methodological Ideal and Actual Procedure", pp. 25–37.

when ideas of those properties are assumed to be part of what we are endowed with by a benign, all-powerful God.

It is the metaphysical application of the precept that the complex is to be understood in the light of the simple that yields the starting point of philosophy in general, and metaphysics in particular. But it is the precept that the complex is to be understood in the light of the simple that is in a sense more basic still. There is no reason to think that conformity to that precept is less important to *scientia* than, or even wholly distinguishable from, the unshakeability of a given conclusion of a process of "deduction".[7]

## IV Metaphysics and the Other Sciences

If *scientia* is a matter of something metaphysics has in common with other sciences, rather than something distinctively metaphysical, there is a sense in which metaphysics is only one science among others, not the pre-eminent Cartesian science. There are also signs that Descartes thinks that metaphysical topics are inappropriate to pursue for long, and should divert people from the other sciences for at most a short time. In Part Six of the *Discourse* and again in the Preface to the French edition of the *Principles*, he shows a marked preference for the practical sciences over the speculative, and claims to be guided even in the speculative parts of his scientific writings by "the law which obliges us to do all that is in our power to secure the general welfare of mankind" (AT VI 61; CSM I 142). So even the non-metaphysical theoretical sciences are supposed to draw their value from their contribution to the mastery of nature for human benefit.

A passage in which Descartes seems to cast doubt on the wholesomeness of metaphysical reflection comes in the letter to Princess Elizabeth of 28 June 1643:

> I believe that it is very necessary to have properly understood, once in a lifetime, the principles of metaphysics, since they are what gives us the knowledge of God and of our soul. But I think it would also be very harmful to occupy one's intellect frequently in meditating upon them, since this would impede it from devoting itself to the functions of the imagination and the senses. I think the best thing is to content oneself with keeping in one's memory and belief the conclusions which one has drawn from them, and then employ the rest of one's study in thoughts in which the intellect co-operates with the imagination and the senses (AT III 695; CSM III 228).[8]

Earlier in the same letter, Descartes has made it clear that he takes his own advice, as in his own studies, he says, he spends very little time—no more than "a few hours a year"—in thoughts which exercise the intellect alone (AT III 692–3; CSM III 227). The solution of specific physics problems and the design of machines helpful to

---

[7] It is possible that the insistence on intuition of simples and deduction from simples is the most enduring of the metascientific claims in the Cartesian corpus. Seeing this has been made harder by misinterpretations of the *Regulae*.

[8] A parallel passage is to be found in *The Conversation with Burman*, J. Cottingham, ed. (Oxford: The Clarendon Press, 1976), p. 30.

human beings occupies the full range of human faculties; and so in a sense they are to be preferred to metaphysics.

The subordination of sciences exercising the intellect and imagination alone to sciences and occupations that exercise intellect, senses *and* imagination reflects Descartes's philosophical anthropology. Human beings are not intellects alone, and it does not come naturally or easily to them to detach themselves from the senses and imagination. Nor is doing so useful outside the once-in-life-time task of discovering the first principles of science, or that of reducing the influence of bad but influential scientific principles—in Descartes' day Aristotelian ones.

What the speculative sciences, including metaphysics, are subordinated to in particular is *medicine* on the one hand and *morals* on the other. The argument for taking medicine to be the pre-eminent science is that it is the science of the "maintenance of health, which is undoubtedly the chief good and the foundation of all the other goods of this life" (AT VI 62; CSM I 143). Or so we are told in Part VI of the *Discourse*. In the Preface to the French edition of the *Principles*, medicine is put alongside mechanics (the science of useful technology) and morals among the fruit-bearing branches of the tree of philosophy. If medicine prolongs life, morals helps to perfect the union of mind and body in the light of an understanding of the perfections that mind and body are capable of. Metaphysics is not irrelevant to morals or medicine, for it is metaphysics that teaches us to distinguish the mental from the physical and that introduces to the perfections of the mind exemplified by God. It is also metaphysics that helps to distinguish the life conserved by medicine from the conservation of the soul.

But metaphysics does not itself deliver human benefit even when it delivers us from the doubt of Meditation One. For the *Meditations* applies its therapy mainly to the intellect that the doubt helps to detach from the body and the senses, and the body and the senses are not detachable parts of human beings: they have to be involved in anything that delivers *human* benefit. Metaphysics is pre-eminent in both being the fundamental kind of *scientia* and in being the science that transmits its certainty to others. It is not the pre-eminent science in other respects; and there are particular sciences and particular small scale scientific results that are far removed from metaphysics. Descartes's philosophy of science accommodates a whole variety of sciences, and his logic accommodates *scientia* with many subject matters.

# *Scientia* and Self-knowledge in Descartes

**Nicholas Jolley**

In *The Search after Truth* and related writings Malebranche strongly criticizes Descartes's thesis that the nature of the mind is better known than the nature of body. In opposition to his mentor Malebranche maintains that whereas we have a clear idea of body, we have no such idea of the mind; we know the mind only by consciousness or internal sensation. In the last twenty years or so Malebranche's critique of Descartes in this area has attracted a good deal of mostly favorable attention, and especially if we stand back a little from the texts, it is not difficult to see why.[1] Descartes gave the world a science of body which is a recognizable ancestor of Newtonian physics; even if Descartes's own physics was seriously flawed, he was right in thinking that a science of the physical world was possible. By contrast, Descartes produced no comparable science of the mind. In this area all Descartes can offer, it seems, is the kind of rational psychology the weaknesses and illusions of which were devastatingly exposed by Kant in the Transcendental Dialectic of the *Critique of Pure Reason*. Indeed, it has been plausibly argued that Malebranche's critique of Descartes is a precursor of Kant's demolition work. Whatever his own intentions may have been Malebranche was in effect engaged in undermining rational psychology from within.[2]

The view that Malebranche has the better of the debate with Descartes has not gone unchallenged; indeed, recently there have been signs of a backlash. Although, to my knowledge, no one has yet sought to rehabilitate Descartes's thesis that the nature of the mind is better known than the nature of body, some scholars have

N. Jolley (✉)
University of California, Irvine
e-mail: sjolley@uci.edu

[1] See C.J. McCracken, *Malebranche and British Philosophy* (Oxford: Clarendon, 1983), pp. 76–81; T. Schmaltz, *Malebranche's Theory of the Soul: A Cartesian Interpretation* (New York: Oxford University Press, 1996), N. Jolley, *The Light of the Soul: Theories of Ideas in Leibniz, Maleranche, and Descartes* (Oxford: Clarendon, 1990), and "Malebranche on the Soul," *The Cambridge Companion to Malebranche*, ed. S. Nadler (New York: Cambridge University Press), pp. 31–58.

[2] A. Pyle, *Malebranche* (New York and London: Routledge, 2003), p. 186.

T. Sorell et al. (eds.), *Scientia in Early Modern Philosophy*, Studies in History and Philosophy of Science 24, DOI 10.1007/978-90-481-3077-1_6,

argued that Descartes was at least entitled to claim epistemic parity in this area.[3] Nolan and Whipple, in particular, are even prepared to defend this claim with reference to the concept of *scientia* that ultimately derives from Aristotle. On their view Descartes claims—justifiably on his principles—that he had achieved *scientia* with regard to mind as well as body.

My aim in this paper is not to rehearse the case for Malebranche's critique of Descartes but rather to challenge the understanding of Descartes's position on which such recent defenses of Descartes rely. The idea of framing the discussion in terms of the concept of *scientia* is a good one, but I believe that the interpretation offered by Nolan and Whipple, and to some extent LoLordo, cannot be sustained. In the first part of the paper I argue that Nolan and Whipple give a mistaken account of Cartesian *scientia*; contrary to their claims, the Cartesian concept of *scientia* is a strong one that retains more of its Aristotelian connotations than they allow. In the second and third parts of the paper I argue that there is no clear evidence that, for Descartes, the conditions for *scientia* are satisfied by his account of mind; moreover, it is a mistake to lift Descartes's claims about self-knowledge out of context and treat them as final results of the system. In the final section of the paper I briefly examine the issue of whether, for Descartes, a *scientia* of mind is even possible. Throughout the paper I focus on Descartes's position in the *Meditations* and related writings and ignore possibly complicating factors introduced by his final philosophical work, the *Passions of the Soul.*[4]

## I Descartes and *Scientia*

The concept of *scientia* derives from the Aristotelian tradition, and whatever else is controversial, one thing is surely clear: though Descartes retains the term "*scientia*" for a particularly valuable and fruitful kind of knowledge, he does not retain the Aristotelian conception in its entirety. To understand the nature of Descartes's break with Aristotle, consider the key components of traditional Aristotelian *scientia* as they emerge from this helpful summary given by Pauline Phemister:

> In the *Posterior Analytics* Aristotle stipulates that scientific knowledge is always knowledge of what is universally true and it proceeds by necessary propositions...Items of scientific knowledge can be demonstrated by syllogistic deductions from true premises which are a priori to us and better known than the conclusion and which contain within them the 'cause'

---

[3] L. Nolan and J. Whipple, "Self-Knowledge in Descartes and Malebranche," *Journal of the History of Philosophy* XLIII (2005), 55–82; A. LoLordo, "Descartes and Malebranche on thought, Sensation, and the Nature of the Mind," *Journal of the History of Philosophy* XLIII (2005), 387–402. LoLordo revealingly remarks that "not much in Descartes's system depends on our having better knowledge of the mind than of he body" (390).

[4] For the issue of Descartes's scientific ambitions for *The Passions of the Soul,* see T. Sorell, "Morals and Modernity in Descartes," *The Rise of Modern Philosophy: The Tension between the New and Traditional Philosophies from Machiavelli to Leibniz,* ed T. Sorell (Oxford: Clarendon Press, 1993), pp. 273–88.

> or explanation of the conclusion. . .This procedure enables us to understand the thing which is to be demonstrated, for it shows why the thing is the way it is and could not possibly be otherwise. We have knowledge of a thing on this model when we know its necessary cause. Knowledge of the cause is provided by demonstration of the fact to be explained.[5]

Such a conception of *scientia* may have held the stage throughout the medieval period, but it was increasingly challenged in the age of the Scientific Revolution.

How much of the Aristotelian conception of *scientia* Descartes rejects may be controversial, but certain things are surely not in doubt. In the first place, Descartes rejects the thesis that *scientia* necessarily involves demonstrating effects from causes. As Hacking says, the method employed by the scientists and natural philosophers of the Scientific Revolution tended to be hypothetico-deductive; practising scientists were increasingly engaged in postulating causes to explain effects, and then deriving test implications from the hypotheses. Descartes may have tried to convince critics that this method too was a kind of demonstration, but it is clearly not demonstration in the traditional Aristotelian sense.[6] Secondly, Descartes of course cannot accept that *scientia* involves syllogistic inference; no philosopher in the early modern period is more famous than Descartes for his hostility to the syllogism. Perhaps misunderstanding its purely expository role in demonstration for Aristotle, Descartes objects that the syllogism is useless as an instrument of discovery.[7] And there were other grounds for dethroning the syllogism from the prominent position it had held for Aristotle. It had never been plausible to claim that Euclidean proofs were syllogistic in form, yet in the early modern period such proofs were widely regarded as paradigm examples of demonstration.

Descartes may have had more distinctively philosophical reasons of his own for rejecting other features of Aristotelian *scientia*. Consider Aristotle's insistence that *scientia* is of universal and necessary truths. Whether Descartes can accept the universality requirement is surely put in question by the *cogito*: if the *cogito* is indeed an ingredient of *scientia*, then *scientia* will include at least some singular propositions. And whether Descartes can accept the necessity requirement is a more interesting question with wider ramifications. The issue is highly controversial, but arguably Descartes is prevented by his doctrine of the creation of the eternal truths from subscribing to this requirement. When Descartes insists on the dependence of the eternal truths on the divine will, he can be read as denying that there are, strictly speaking, any necessary truths. Descartes seems to suggest such a reading when he writes to Mersenne of "the mathematical truths which you call eternal" (15 April 1630, AT I 145: CSMK III 23); he thereby seems to distance himself from the thesis that there are any eternal, that is, necessary truths. Some commentators of course

---

[5]P. Phemister, "Locke, Sergeant, and Scientific Method," *The Rise of Modern Philosophy*, ed. Sorell, p. 232.

[6]See I. Hacking, "Leibniz and Descartes: Proof and Eternal Truths," *Rationalism, Empiricism, and Idealism*, ed. A. Kenny (Oxford: Clarendon, 1986), p. 55.

[7]On Descartes's critique of the syllogism, see S. Gaukroger, *Cartesian Logic* (Oxford: Clarendon, 1989). Gaukroger emphasizes Descartes's criticism that the syllogism is useless as an instrument of discovery.

have read Descartes as advancing a weaker thesis: although there are necessary truths, they are not necessarily necessary.[8] At first sight such an interpretation is encouraged by a passage from a letter to Mesland:

> And even if God has willed that some truths should be necessary, this does not mean that he willed them necessarily; for it is one thing to will that they be necessary, and quite another to will this necessarily, or to be necessitated to will it. (2 May 1644, AT IV 118–19: CSMK III 235)

But as I have argued elsewhere, even here Descartes stops significantly short of endorsing the thesis that there are necessary truths.[9] He appears rather to be arguing concessively: even if it is granted that there are necessary truths, it does not follow that they are necessarily necessary. The stronger reading of Descartes's doctrine is in line with what we may regard as one of the main morals of the Cartesian revolution in philosophy—the shift away from modal claims to epistemic ones.

Descartes, then, has reasons stemming both from the science of his time and from his own philosophy for not endorsing the full traditional Aristotelian conception of *scientia*. But what conception of *scientia* does he put in the place of the traditional Aristotelian one? Nolan and Whipple have recently answered this question by saying: "Descartes consistently characterizes *scientia* as a variety of certainty that is grounded in knowledge of the existence and nature of God who guarantees that our intellectual faculty cannot but tend towards the truth."[10] Such an interpretation seems to be encouraged by the enquirer's reflection on his situation at the end of the Fifth Meditation. "Thus I see plainly that the certainty and truth of all knowledge (*scientiae*) depends uniquely on my knowledge of the true God, to such an extent that I was incapable of perfect knowledge about anything else until I knew him" (AT VII 71: CSM II 49).

But this account of Cartesian *scientia* is, I believe, open to challenge. In the first place, it can be criticized on textual grounds. According to Nolan and Whipple, Descartes *consistently* invokes the divine guarantee in his characterizations of *scientia*. But this textual claim is not strictly accurate. In the unfinished French work *The Search for Truth*, for instance, Descartes's spokesman, Eudoxus, speaks of acquiring "a body of knowledge *(doctrine)* which was firm and certain enough to deserve the name 'science' (*science*)" (AT X 513: CSM II 408). Here there is no explicit mention of the divine guarantee. But secondly, and more importantly, Nolan and Whipple omit a condition that is stated in the quotation from the *Search*: *scientia* is a body of knowledge as opposed, say, to a set of isolated intuitions. This insistence on *scientia* as a body of knowledge is one of the features that arguably distinguish *scientia* for Descartes from mere cognition (*cognitio*); it also constitutes one important remaining link with the Aristotelian tradition.

---

[8]See E. Curley, "Descartes on the Creation of the Eternal Truths," *Philosophical Review* 93 (1984), 569–97. Cf. Hacking, "Leibniz and Descartes: Proof and Eternal Truths," pp. 52–3.

[9]Jolley, *Light of the Soul*, p. 51. Cf. H. Frankfurt, "Descartes on the Creation of the Eternal Truths," *Philosophical Review* 86 (1977), 36–57.

[10]Nolan and Whipple, "Self-Knowledge in Descartes and Malebranche," 63.

One major weakness of the account offered by Nolan and Whipple is that, while emphasizing the importance of the divine guarantee, it fails to grasp its purpose; that is, it fails to recognize that God's guarantee allows our knowledge to become systematic. The point has been admirably made by John Cottingham in his introduction to his edition of the *Conversation with Burman*:

> The need for God in Descartes's theory of knowledge, and the sense in which all knowledge can be said to depend on him, now begins to emerge. For although we can have some knowledge without God (the knowledge of epistemically self-guaranteeing propositions), such knowledge would never, so to speak, get us anywhere. It would last only as long as the relevant proposition, or set of propositions, was actually being attended to. . ..Once we have arrived at the proposition that God exists and is not a deceiver, then at last the possibility of developing a *systematic body* of knowledge becomes available. (CB xxxi–xxxii)

Notice that, on this view, the divine guarantee may not be built into very definition of *scientia;* it may be rather that which explains essential features of *scientia* such as firmness and systematicity. As we have seen, the divine guarantee is omitted from the definition of *scientia* in *The Search For Truth.* But whether we hold that the divine guarantee is built into the very definition of *scientia* matters little: the important point is to understand its role or purpose in Descartes's epistemology.

Before we leave the issue of the nature of Cartesian *scientia,* we should notice one further claim that Nolan and Whipple press: this is the alleged anti-formalism of Descartes's conception of *scientia.* Following the lead given by Hacking and others, Nolan and Whipple insist that it is a mistake to think of *scientia* as a systematic body of knowledge on the traditional Aristotelian model. *Scientia,* for Descartes, is not an axiomatic system "in which the various parts of knowledge bear complex entailment relations to one another".[11] On their view, Descartes is most interested in "unveiling the contents of our clear and distinct ideas and thereby attaining knowledge," and this "has nothing to do with deducing theorems from axioms in the traditional sense, and everything to do with removing prejudices so that these contents can be immediately intuited".[12]

Nolan and Whipple make some useful points, but they seem to me to be in danger of conflating two questions. That is, they seem to run together the issue of whether *scientia* is necessarily a systematic body of knowledge with the issue of whether it must be conceived as a formal axiomatic system. Now there is no doubt that Descartes is generally hostile to traditional formal logic; in particular, as we have seen, he despises the syllogism because of its uselessness as an instrument of discovery. But from the fact that *scientia* need not be a formal system for Descartes, it does not follow that it is not essentially a systematic body of knowledge; even if the truths in the system are deductively linked, it is still possible to give an account of such deduction that is non-formal. Moreover, even if, for Descartes, formal deduction plays no role in the discovery of new truths, it does not follow that it plays no role in displaying the systematic structure of such knowledge. And as Gaukroger

---

[11] Ibid., p. 61.

[12] Ibid., p. 62.

says, this is the role that it plays in such a work as *The Principles of Philosophy.*[13] Thus the issue of Descartes's anti-formalism seems irrelevant, or marginal at best, to the proper understanding of his concept of *scientia*.

An underlying weakness of the Nolan and Whipple account is the exaggerated and misleading insistence on Cartesian therapy. Like other commentators they stress that Descartes seeks to offer a cure in the *Meditations* for the prejudices and preconceived opinions which go back to childhood and to replace them with the clear and distinct ideas of the intellect. Descartes's interest in such epistemological therapy may be real, but it should not lead us to underestimate his ambition to be the new Aristotle. That is, Descartes is not just interested in showing us how to achieve exquisite states of certainty about our own existence and the existence of God; he is interested, surely much more, in developing a new and true science of the physical world to replace the discredited Aristotelian one. As Cottingham says, Descartes wants to show us how our knowledge can get somewhere. And we should never forget that on the first page of the First Meditation—the work which is cited as primary evidence of Descartes's concern with therapy—the enquirer explains his real ambitions and the goal of his whole enterprise: demolishing everything completely and starting again on new foundations is seen to be necessary "if I wanted to establish anything at all in the sciences (*scientiis*) that was stable and likely to last" (AT VII 17: CSM II 12). Descartes is more interested in being a natural philosopher than a psychotherapist.

## II *Scientia* and the Mind: The Second Meditation

*Scientia*, for Descartes, is thus a systematic body of knowledge that is firm and certain and that is made possible by the divine guarantee. Now it is uncontroversial that Descartes supposed his physics to satisfy the conditions for *scientia*; as we have seen, those features of the traditional Aristotelian conception that were problematic for the new science have been quietly (or not so quietly) discarded by Descartes. To say that Cartesian physics satisfies the definition of *scientia* is not to say that the interpretation of the physics raises no problems. Commentators have debated such issues as the role of experience—that is, observation and experiment—in Cartesian science and the extent to which it is supposed to be *a priori*.[14] One may also wonder how literally we are supposed to take Descartes's claim in correspondence that "my entire physics is nothing but geometry" (27 July 1638, AT II 268: CSMK III 119).

---

[13] Gaukroger, *Cartesian Logic*, p. 116.

[14] For important discussions of Descartes's science, see D. Garber, *Descartes's Metaphysical Physics* (Chicago: University of Chicago Press, 1992) and "Descartes' Physics," *The Cambridge Companion to Descartes*, ed. J. Cottingham (Cambridge and New York: Cambridge University Press, 1992), pp. 286–334; D. Clarke, *Descartes' Philosophy of Science* (Manchester: Manchester University Press, 1982), *Occult Powers and Hypotheses: Cartesian Natural Philosophy Under Louis XIV* (Oxford: Clarendon, 1989) and "Descartes' Philosophy of Science and the Scientific Revolution," *Cambridge Companion to Descartes*, ed. Cottingham, pp. 258–85.

But it is beside my purpose here to enter into such controversies. I shall simply take it for granted that Cartesian physics is supposed to be a paradigm instance of *scientia*. Of course, as I indicated in the introduction, we know that Cartesian physics is seriously flawed; Newton himself annotated his copy of Descartes's *Principles of Philosophy* by writing the word "error" in the margins over and over again. But the fact that we now know that Descartes's physics is not true—that in words attributed to Pascal it provides merely a romance of nature[15]—is irrelevant to the issue of how Descartes viewed his achievement.

There is thus supposed to be Cartesian *scientia* of the physical world; is there also be supposed to be a Cartesian *scientia* of the human mind? It is not difficult to see how one could come to think that there must be. Descartes is famous, or notorious, for his thesis that the mind is better known than body, and since Descartes clearly holds that he has developed a *scientia* of body, it may well seem that he is committed to the thesis that there is a *scientia* of the mind. Certainly Nolan and Whipple take such a view, for they write that "our knowledge of the mind's nature is at least on a par with our knowledge of corporeal nature",[16] and as we have seen, they offer a rather minimal characterization of Cartesian *scientia* as certain knowledge that is underwritten by the divine guarantee.[17] But it is, I suggest, a mistake to suppose that the thesis of the Second Meditation is relevant to the issue of whether, for Descartes, there is a *scientia* with regard to the nature of the mind. As we shall see, Descartes is concerned here, not with *scientia*, but with an inferior or at least less fruitful kind of knowledge.

The fact that the thesis that the mind is better known than body is introduced and defended in the Second Meditation is important, for even though it is defended elsewhere (for example, in the Fifth Replies), it needs to be understood, not as a thesis within the final system, but rather in terms of the stage of the journey which the enquirer has reached. Remember that the later part of the Second Meditation has the goal of combatting twin empiricist prejudices—the view that we know bodies best of all and the view that bodies are known through the senses. The enquirer is engaged in rehearsing arguments to rid himself of these prejudices once and for all. Moreover, we should not expect Descartes to say that the enquirer is in possession of a *scientia* of the mind. For at this stage the enquirer still lacks the divine guarantee; and whether or not the divine guarantee is built into the very definition of "scientia", it is in some sense a necessary condition of such knowledge. Of course

---

[15]B. Pascal, *Pénsees*, ed. A.J. Krailsheimer (Harmondsworth: Penguin, 1966), p. 356

[16]Nolan and Whipple, "Self-Knowledge in Descartes and Malebranche," 56.

[17]My concern in this article is primarily with Descartes's view of his project, not with his achievement, but it is worth noting that Nolan and Whipple offer characterizations of Descartes's procedure which are both misleading and too generous. For instance, they write not only of the "*res cogitans* proof" in the Second Meditation but also of his discovery of further properties of the mind a priori from our innate idea of the self (Nolan and Whipple, "Self-Knowledge in Descartes and Malebranche," 65–6.) Such descriptions of Descartes's procedure appear unwarranted. In the Second Meditation, for instance, Descartes seems to me not to discover properties of the mind a priori but simply to appeal to the data of introspection.

it may be objected that there is logical space for a distinction between not having the divine guarantee at all and having the divine guarantee but not knowing that one has it. It might then be said that Descartes's enquirer is in the second position and not the first. But as the discussion of the atheist geometer shows, it seems that, for Descartes, having the divine guarantee essentially involves knowing that one has it—that is, knowing that God exists and would not deceive us with regard to our clearest intellectual intuitions.

We should adopt the same approach, I believe, to the Fifth Replies. Gassendi famously and rather naturally objects that Descartes may well have established that the existence of his mind is more certain than the existence of body, but he has not succeeded in establishing the more important thesis that the nature of the mind is better known than the nature of body (AT VII 275: CSM II 192). Descartes responds to this line of objection in his own voice, but his response, I suggest, should still be seen as relativised to the stage that the enquirer has reached on his philosophical journey. It may be objected that if this is what Descartes is doing, it is strange that he does not make the point explicitly; moreover, his response is most naturally read as a defense of a result in the final system. But such objections are arguably insensitive to the text. First, consider Descartes's impatient response to Gassendi's demand for a chemical investigation of the mind: "Nor do I see what more you expect *here*" (AT VII 359: CSM II 248; emphasis added), where the "here" is a reminder that the enquirer's argument should be understood in context. Moreover, we should notice that a little earlier Descartes has given a much more explicit reminder to Gassendi that his arguments should be understood in context—that is, in terms of the stage in his philosophical journey that the enquirer has reached in the Second Meditation. Descartes reminds Gassendi that he had said that insofar as he knew himself he was nothing other than a thinking thing, and significantly adds: "This is all that I asserted in the Second Meditation" (AT VII 355: CSM II 245). It would be strange indeed if a few pages later Descartes were to forget that the issue at hand is simply what the enquirer has come to understand at this point in the Second Meditation.

The fact that in the Second Meditation and related writings Descartes is concerned with something less than *scientia* is confirmed by the evidence of terminology. In all his statements of the thesis that mind is better known than body Descartes consistently avoids using the term *scientia.* The title of the Second Meditation informs us simply: mind is *notior* than body (AT VII 23: CSM II 16), and in the body of the meditation itself he summarizes his result by saying: "aperte cognosco nihil facilius aut evidentius mente posse a me percipi" (AT VII 34: CSM II 22–3). The same avoidance of any reference to *scientia* is apparent in the Fifth Replies where Descartes defends his thesis against Gassendi by reference to the principle that we know something better the more attributes we know of it. The interpretation of this principle is not our present business, and we shall return to it; here our concern is with the fact that Descartes uses terms like *cognitio* and *cognoscere*, not *scientia*:

> But as for me, I have never thought that anything more is required to reveal a substance than its various attributes, thus the more attributes of a substance we know (*cognoscamus*), the more perfectly we understand its nature...The clear inference from this is that more

> attributes are known (*cognosci*) in the case of our mind than in the case of anything else. For no matter how many attributes are recognized (*cognoscuntur*) in any given thing, we can always list a corresponding number of attributes in the mind which it has in virtue of knowing the thing: and hence the nature of the mind is the one that is known best of all (*notissima*). (AT VII 360: CSM II 249; translation modified)

The same avoidance of the term "*scientia*" is found in the corresponding passage from the *Principles of Philosophy,* the marginal summary of which is: "Quomodo mens nostra notior sit quam corpus":

> In order to realize that our mind is known (*cognosci*) not simply prior to and more certainly (*certius*) but also more evidently than body, we should notice something very well known by the natural light: nothingness possesses no attributes. It follows that whenever we find some attributes or qualities, there is necessarily some thing or substance to be found for them to belong to; and the more attributes we discover in the same thing or substance, the more clearly do we know (*cognoscere*) that substance. Now we find more attributes in our mind than in anything else, as is manifest from the fact that whatever enables us to know (*cognoscamus*) anything else cannot but lead us to a much surer knowledge (*cognitionem*) of our own mind. (AT VIIIA 8: CSM I 196; translation modified)

It is tempting to emphasize that Descartes is not talking about *scientia* in such passages by translating "*cognitio*" and its cognates by "acquaintance"; thus we might read Descartes as saying that we have a much surer acquaintance with our mind than anything else. It is true that such a translation might on occasion sound rather odd; it would give the reader a jolt to be told that "the more attributes of a given substance we are acquainted with, the more perfectly we understand its nature" (AT VII 360: CSM II 249). To my ear at least, the passage would be less jolting if "*perfectius*" were translated as "more completely", as is indeed quite acceptable. And as we shall see, there are reasons why we should not be too worried if this criterion of complete understanding sounds naive. But it would be a mistake to suggest that "*cognitio*" can always be rendered as "acquaintance" if acquaintance is taken to involve non-propositional knowledge. Consider, for instance, Descartes's well-known discussion of the problem of the atheist geometer where he makes one of his sharpest distinctions between *cognitio* and *scientia*. Descartes writes of the atheist's *cognitio* that the three angles of a triangle are equal to two right angles; here the *cognitio* is obviously propositionally structured. Cottingham, Stoothoff, and Murdoch are probably right to settle for the word "awareness" in their translation:

> The fact that an atheist can 'clearly know that the three angles of a triangle are equal to two right angles', I do not dispute. But I maintain that this awareness (*cognitio*) of his is not true knowledge (*scientia*), since no act of awareness that can be rendered doubtful seems fit to be called knowledge (*scientia*). (AT VII 141: CSM II 101; translation modified)

Thus I shall not insist on the claim that, in the passages from the Fifth Replies and the *Principles*, "acquaintance" may be a better translation than "knowledge". The important point is that Descartes constantly uses terms which, unlike *scientia*, have no connotations of systematic knowledge.

Understanding that Descartes's reply to Gassendi is relativized to the enquirer's stage in the *Meditations* may help us to meet a well-known critique of Descartes's argumentative strategy in the Fifth Replies. Recall that Descartes famously defends the thesis that mind is better known than body by appealing to the principle that the more properties of a substance we know, the more perfectly we understand its nature. Commentators have objected that this criterion of perfect knowledge is a simplistically quantitative one which is at odds with the overall tenor of Descartes's philosophical system. Margaret Wilson makes the point well:

> For the predominant theme in his writings on knowledge of nature is that perfect comprehension of material substance is obtained not by lengthening the list of properties (as Bacon's program, for instance, demanded), but by providing an account of the extension, figure, and motion of body's internal parts.[18]

Wilson further objects that Descartes's application of the criterion in the Fifth Replies leads him into an inconsistency with the main theme of the wax sub-meditation. For in explaining the application of the criterion in response to Gassendi, Descartes gives a long list of sensory judgments about the piece of wax; that is, he lists sensible qualities such as whiteness and hardness which are identified through judgments based on sensory intake. By contrast, in the wax meditation itself Descartes had located perfect understanding in "an intellectual perception of the essence of body".[19] Thus, according to Wilson, Descartes seems to have forgotten the very moral of the wax meditation.

It is possible to reply to Wilson's first objection here by saying that there is indeed a sense in which the quantitative criterion of perfect knowledge is not the criterion that Descartes endorses when expounding his system. But it does not follow from this that Descartes is at fault for not invoking the more sophisticated criterion in his response to Gassendi. For once we recognize that Descartes's reply to Gassendi is relativized to the enquirer's stage in the Second Meditation, we can see that he is stating a criterion that is appropriate to the enquirer's level of philosophical enlightenment. Moreover, the criterion in question in the Fifth Replies is a criterion not of *scientia* but of *cognitio*. Thus to the objection that Descartes does nothing to tell Gassendi that the criterion in question is a simplistic one, we may concede the point, while adding a qualification that effectively draws its sting: from the fact that it is a simplistic criterion of perfect *scientia* it does not follow that it is a simplistic criterion of perfect *cognitio*. These two responses can, I think, be combined. The enquirer, at the stage of philosophical enlightenment he has reached in the Second Meditation, does not yet grasp the nature of *scientia* and its relation to *cognitio*. Wilson has essentially made the mistake of confusing the criteria of perfect *scientia* and the criteria of perfect *cognitio*.

Descartes's appeal to the apparently simplistic criterion can thus be defended. What of Wilson's objection that the reply to Gassendi misrepresents the moral of the very meditation that it is supposed to be explaining and defending? According to Wilson, as we have seen, in the original discussion of the piece of wax Descartes identifies perfect comprehension with intellectual perception of the essence of

---

[18]M.D. Wilson, *Descartes* (London: Routledge, 1978), pp. 96–7.

[19]Ibid., p. 97.

body. At this point we stumble on the central problems of interpreting the wax meditation—a passage that has provoked widely different readings, and we cannot do them full justice here. But it is arguable that Wilson reads too much into the wax meditation. Notice that her phrase "intellectual perception of the essence of body" in effect combines metaphysical considerations (about the essence of body) and epistemological ones (about how such essences are perceived). But it is worth recalling that one of Descartes' two main aims at this stage is simply to refute the naive empiricist view that bodies are known through the senses. All he needs to establish to that end is the epistemological thesis that bodies are perceived through the intellect; it is the intellect, for example, that grasps that a body can remain the same through an infinity of changes in sensible qualities; at this stage the reality of these qualities need not be called into question. Thus a defense of the core thesis about how bodies are known does not require any claim about the essence of bodies or how this essence is grasped. Wilson's charge that the criterion of perfect knowledge which Descartes states and defends in response to Gassendi is inconsistent with the teaching of the original wax meditation arguably depends on reading back the final results of the system into the Second Meditation where they do not belong.

## III *Scientia* and the Mind: Other Texts

So far I have concentrated on the Second Meditation and related writings because it is here that Descartes introduces and defends the thesis that mind is better known than body. And though, as we have seen, they have not been prepared to defend the letter of his thesis, commentators who have sought to defend Descartes against Malebranche's critique have also emphasized those texts, or at least not discounted them. But it may be objected that it is misguided to focus on such texts to the exclusion of others that are more relevant; for what is at issue is the set of commitments of Descartes's completed system. Thus we must look to the Sixth Meditation and the relevant sections in the *Principles of Philosophy* where Descartes is expounding the final system.

It is indeed necessary to look beyond the Second Meditation since, as I have emphasized, this represents only a stage on the enquirer's journey to full philosophical enlightenment. It cannot be denied that the enquirer makes epistemic progress between the Second and Sixth Meditations; in particular, he discovers that he has a divine guarantee for his clearest intellectual intuitions. But with regard to the issue of *scientia* the picture that emerges when we take a broader view is not significantly different. Even when the enquirer is nearing the end of his philosophical journey Descartes still avoids saying that he has *scientia* of the mind or can achieve it. In the Sixth Meditation the enquirer discovers that he has come to know God and himself better (*melius nosse*) (AT VII 77: CSM II 54), and that he can achieve a clear and distinct idea of the mind and its real distinction from the body (AT VII 78: CSM II 54), but even when underwritten by the divine guarantee clear and distinct perception does not entail *scientia*. The same pattern is repeated in the corresponding sections of the *Principles of Philosophy* where Descartes is magisterially expounding the results of his system. Descartes explains how substances are known:

> A substance is known *(cognoscitur)* through any attribute at all; but each substance has one principal property which constitutes its nature and essence, and to which all its other properties are referred. Thus extension in length, breadth and depth constitutes the nature of corporeal substance; and thought constitutes the nature of thinking substance. (AT VIIIA 25: CSM I 210; translation modified)

Thus what is at issue here is *cognitio*, not *scientia*.

The claim that, for Descartes, there is no *scientia* of the mind may encounter some resistance. Critics are likely to point to the end of the Fifth Meditation for evidence to the contrary. Here it might seem that Descartes is clearly committed to the claim that we can achieve *scientia* with regard to the nature of the mind. Consider not only the well-known statement at the end of the Fifth Meditation that the divine guarantee is necessary for the achievement of *scientia* but also the way in which the enquirer then follows it up. Here, by using the term "knowledge" throughout, the standard English translation by Cottingham, Stoothoff and Murdoch certainly encourages the idea that *scientia* is at issue in both sentences:

> Thus I see plainly that the certainty and truth of all knowledge depends uniquely on my knowledge of the true God, to such an extent that I was incapable of perfect knowledge about anything else until I knew him. And now it is possible for me to achieve full and certain knowledge of countless matters, both concerning God himself and other things whose nature is intellectual, and also concerning the whole of that corporeal nature which is the subject matter of pure mathematics. (AT VII 71: CSM II 49)

On the plausible assumption that human minds are included among the intellectual things here, it would indeed seem from this translation that Descartes is talking about the prospects for *scientia* throughout. But the translation is arguably misleading: consulting the Latin text shows that there is a switch in terminology, for the second sentence reads:

> Jam vero innumera, tum de ipso Deo aliisque rebus intellectualibus, tum etiam de omni illa natura corporea, quae est purae Matheseos objectum, mihi plane nota & certa esse possunt.

Descartes thus stops short of using the term *scientia* to characterize the knowledge of God, intellectual things, and corporeal nature that the divine guarantee makes possible.

It is natural to object that in this passage, whatever we make of the switch in terminology, Descartes clearly seems to place God, intellectual things, and corporeal nature on the same epistemic footing. But this shows less than one might expect. For Descartes's point may be a rather limited one: once we are in possession of the divine guarantee, an obstacle to the attainment of *scientia* is removed across the board; whether the subject matter of our enquiry is God, mind or body, we no longer need to worry that our nature is defective and that we may be systematically deceived with regard to our clearest intellectual intuitions. To that extent God, intellectual things, and corporeal nature are indeed epistemically on a par. But to say this is not to say that *scientia* is equally attainable in all fields; for some of these areas of enquiry may be such that they do not lend themselves to the satisfaction of the systematicity condition. The possession of divinely guaranteed intuitions or acts of awareness is a necessary condition for *scientia*, but is not a sufficient one.

The claim that, for Descartes, there is no *scientia* of the mind is also likely to encounter resistance from a related quarter. It is beyond dispute that one of the results of Descartes's system is the discovery of the essence of the mind: that essence is constituted by thought or *cogitatio*. And it may be supposed that if Descartes holds that we can know the essence of mind, he must surely hold that we can achieve *scientia* in this regard. Here the idea is that knowing the essence of *x* necessarily involves the ability to demonstrate non-trivial properties of *x* in a way that yields the systematic knowledge which constitutes *scientia*. Such a reading is encouraged by Descartes's famous discussion of true and immutable natures in the Fifth Meditation:

> When, for example, I imagine a triangle, even if perhaps no such figure exists, or has ever existed, anywhere outside my thought, there is still a determinate nature, or essence, or form of the triangle which is immutable and eternal, and not invented by me or dependent on my mind. This is clear from the fact that various properties can be demonstrated of the triangle, for example that its three angles equal two right angles, that its greatest side subtends its greatest angle, and the like. (AT VII 64: CSM II 44–5)

To know the essence of the triangle thus involves the ability to demonstrate the properties that Euclid proves in the *Elements*. And no one would doubt that Euclidean geometry has the systematic nature required for *scientia*.

But the famous discussion of true and immutable natures is, I believe, misleading with regard to Descartes's general position; the geometrical case is in no way analogous to the case of the mind. When Descartes says that various interesting properties follow from the essence of the triangle, he is invoking a thick concept of essence that includes not just the definition but the axioms and even postulates of Euclidean geometry; the properties to which he appeals do not follow from the essence of the triangle taken more strictly as the definition. But when Descartes says that we know the essence of the mind, he is not saying that we have epistemic access to an essence in the sense he invokes in the geometrical case. In the case of the mind there is nothing comparable to the axioms and postulates of Euclidean geometry. And in the absence of such further propositions to serve as premises, there is no prospect for the demonstration of non-trivial properties that *scientia* would require.

## IV The Possibility of a *Scientia* of Mind

There is thus no clear evidence that Descartes claims to be in possession of a "*scientia*" of mind. If this conclusion is correct, then it naturally prompts the question whether Descartes is committed to holding that the search for *scientia* in this area is misguided in principle. In favor of this claim we may cite the fact that in the Fifth Replies Descartes criticizes Gassendi for demanding a chemical investigation of the nature of the mind; he seems to suggest that Gassendi's demand is inappropriate as well as question-begging. And it is at least instructive to note a feature of the subsequent controversy between Malebranche and Arnauld; Arnauld, who is in general a reliable proxy for Descartes himself, rebukes Malebranche for demanding the impossible when he criticizes Descartes for mistakenly claiming to be in possession of a clear idea (i.e. *scientia*) of the mind.[20]

It is tempting to mount a very simple argument to show why Descartes might think a *scientia* of the mind is impossible. It might be argued that there are two necessary conditions of the possibility of a *scientia* of the mind that Descartes is unable to satisfy, at least if *scientia* involves systematicity: first, determinism and secondly, complexity of mental structure.[21] These conditions cannot be satisfied in Descartes' philosophy, for Descartes is a libertarian who is committed to the existence of contracausal freedom, and he is a dualist who upholds the doctrine of the simplicity of the soul.

But this argument is too quick: it is vulnerable to two distinct kinds of criticism. One may question not only whether Descartes in fact holds the views that are ascribed to him here, but also whether the allegedly necessary conditions of a *scientia* of the mind are in fact necessary. In the first place, Descartes's position on the issue of free will is controversial; although he has been traditionally read as a libertarian, some recent commentators have argued that there is nothing in the texts which is inconsistent with soft determinism. Secondly, it may be a mistake to suppose that complexity of mental structure can be understood only on a materialist model of the mind; although it is obviously consistent with such a doctrine, it does not seem to entail it. Leibniz indeed offers an instructive example of a philosopher who upholds the simplicity of the soul while also insisting on the complexity of mental structure; for Leibniz, the mind has a complex structure inasmuch as it has an infinity of *petites perceptions* which serve to ground its dispositional properties.[22] Thus it seems possible to grant that complexity is a necessary condition of a *scientia* of the mind while also holding that the complexity is not precluded by a commitment to an immaterialist theory of the mind. Now Descartes of course does not have the Leibnizian doctrine of *petites perceptions*, but he at least shares with Leibniz a commitment to the thesis that the mind has dispositional properties the activation of which results in occurrent mental states. And this may be all that he needs.

Whether Descartes is committed to determinism and complexity of mental structure may be disputed, but even if he is not, it might still be argued that there is room in his philosophy for acknowledging the possibility in principle of a kind of *scientia* of the mind. Once again it is instructive to consider one of Descartes's successors. Malebranche, for example, seems to hold that a *scientia* of the mind

---

[20] A. Arnauld, *On True and False Ideas,* ed. S. Gaukroger (Manchester: Manchester University Press, 1990), Ch. 23, p. 175.

[21] See Pyle, *Malebranche*, p. 188; Nolan and Whipple, "Self-Knowledge in Descartes and Malebranche," 76.

[22] In the Preface to the *New Essays on Human Understanding* Leibniz stresses the parallels between physics and "pneumatology" (RB 56); both sciences postulate unobservables—insensible corpuscles in physics and *petites perceptions* in the case of "pneumatology".

is possible at least in principle while denying both determinism and complexity of mental structure. Such a *scientia* would not of course be a predictive science, but as Malebranche observes, it would involve the ability to know a priori the modifications of which the mind is capable and the true relations between mental states. In other words, such a *scientia* would be closer to geometry than to physics. It is not obvious that Descartes has the resources to rule out the possibility in principle of a *scientia* of the mind conceived on this model. What is clear, however, is that it could not take quite the form that it does in Malebranche. For when Malebranche explicates the possibility at least in principle of a *scientia* of the mind, he does so in terms of an idea of the mind in God which is logically prior to his will; this idea is supposed to be an eternal archetype or blueprint for creation. It is this idea that God has withheld from human beings with the result that we can never achieve *scientia* of the mind. But Descartes's insistence on the strict simplicity of God—a simplicity so strict that there is no distinction between his intellect and his will—precludes him from thinking of the possibility of *scientia* of mind in these Malebranchean terms.

Yet it would be wrong to end by simply emphasizing the distance between Malebranche and Descartes. For the moral of the paper is in a sense that Descartes may be closer to Malebranche on the basic issue than has been realized. Indeed, at least before the *Passions of the Soul,* Descartes can agree with Malebranche that while we possess a science of body we possess no science of the mind. The fact that this point of kinship has not been recognized may arguably be blamed at least in part on Malebranche himself. For while Malebranche was absolutely right to argue that Descartes gives us no *scientia* of the mind, he failed to see that this may not have been Descartes's ambition; in particular, he failed to see that such *scientia* of the mind was not at issue in Descartes's defense of the thesis that the mind is better known than body. To this extent, and to this extent only, those who find fault with Malebranche's critique of Descartes may be justified.[23]

[23] An earlier version of this paper was read to a Philosophy Department colloquium at Rice University. I am grateful to the audience, and to Mark Kulstad in particular, for helpful comments.

# Spinoza's Theory of *Scientia Intuitiva*

**Don Garrett**

Many prominent distinctions involving kinds of knowledge or cognition are dichotomous: a priori or a posteriori, necessary or contingent, analytic or synthetic, conceptual or empirical, certain or probable, self-evident or inferential, general or particular, intellectual or imaginative. In his *Ethics*, however, Spinoza distinguishes and discusses not two but three kinds of cognition [*cognitio*, often translated as 'knowledge'].[1] These three kinds are: (i) *opinion* or *imagination* [*opinio, vel imaginatio*], which is further distinguished into random experience [*experientia vaga*][2] and cognition from signs [*ex signis*]; (ii) *reason* [*ratio*]; and (iii) *scientia intuitiva* [often left in Latin by commentators, but signifying "intuitive knowledge" in a much stricter sense of 'knowledge']. While they are foreshadowed by a similar distinction among "kinds of perception" in two earlier works (the unfinished *Treatise on the Emendation of the Intellect* [TdIE §§ 18–29] and the long-unpublished *Short Treatise on God, Man, and his Well-Being* [KV II.1–2]),[3]Spinoza introduces them into

D. Garrett (✉)
New York University, Department of Philosophy, 5 Washington Place, New York, NY, 10003
e-mail: don.garrett@nyu.edu

[1] The English translation of Spinoza's term *'cognitio'* as 'knowledge' is well entrenched. However, following Jonathan Bennett (1984), I will instead use the cognate term 'cognition', since Spinoza's '*cognitio*' includes within its scope ideas that he characterizes as "inadequate" and "false." In any case, it must be clearly distinguished from knowledge in the much stricter sense of *'scientia'*, which constitutes just one kind of *cognitio*.

[2] For a highly informative discussion of *experientia vaga*, see Gabbey 1996.

[3] All quotations from the *Treatise on the Emendation of the Intellect* [*Tractatus de Intellectus Emendatione*] and the *Short Treatise on God, Man, and His Well-Being* [*Korte Verhandeling van God, de Mensch en deszelfs Welstand*] employ the translation of Edwin Curley in Spinoza 1985, except that 'cognition' replaces 'knowledge' in translating '*cognitio*' and '*scientia intuitiva*' is left untranslated. Citations of the former begin with 'TdIE' and give the paragraph numbers; citations of the latter begin with 'KV' and give the book and chapter numbers. The *Treatise on the Emendation of the Intellect* counts four kinds of perception rather than three because perception from signs and from random experience are counted as two different kinds of perception. The term '*scientia intuitiva*' occurs only in the *Ethics*.

T. Sorell et al. (eds.), *Scientia in Early Modern Philosophy*, Studies in History and Philosophy of Science 24, DOI 10.1007/978-90-481-3077-1_7,

the *Ethics* at the conclusion of a discussion of the "universal notions" that allow us "to perceive many things." He begins by defining them:

> [I]t is clear that we perceive many things and form universal notions:
>
> I. from singular things which have been represented to us through the senses in a way that is mutilated, confused, and without order for the intellect (see P29C); for that reason I have been accustomed to call such perceptions cognition from random experience;
>
> II. from signs, e.g., from the fact that, having heard or read certain words, we recollect things, and form certain ideas of them, which are like them, and through which we imagine the things (P18S). These two ways of regarding things I shall henceforth call cognition of the first kind, opinion or imagination.
>
> III. Finally, from the fact that we have common notions and adequate ideas of the properties of things (see P38C, P39, P39C, and P40). This I shall call reason and the second kind of cognition.
>
> In addition to these two kinds of cognition, there is (as I shall show in what follows) another, third kind, which we shall call *scientia intuitiva*. And this kind of knowing proceeds from an adequate idea of the [NS: formal] essence of certain attributes of God to the adequate cognition of the essence of things.

He then provides an example that he had also used in the earlier works:

> I shall explain all these with one example. Suppose there are three numbers, and the problem is to find a fourth which is to the third as the second is to the first. Merchants do not hesitate to multiply the second by the third, and divide the product by the first, because they have not yet forgotten what they heard from their teacher without any demonstration, or because they have often found this in the simplest numbers, or from the force of the Demonstration of P7 in Bk. VII of Euclid, viz. from the common property of proportionals. But in the simplest numbers none of this is necessary. Given the numbers 1, 2, and 3, no one fails to see what the fourth proportional number is and we see this much more clearly because we infer the fourth number from the ratio which, in one glance, we see the first number to have the second. (E2p40s2)[4]

Spinoza clearly regards the third kind of cognition—the only one on which the *Ethics* bestows the honorific title *'scientia'*—as the best and most desirable of the three; yet in many ways it is also the most puzzling. One question concerns its very nature: What are the essences of attributes and of things on which *scientia intuitiva* depends, and how does it proceed from the former to the latter? A second question concerns its relation to other kinds of cognition: Given that *all* cognition, according to Spinoza, requires an adequate idea of an attribute of God, how does *scientia intuitiva* differ from other kinds of cognition? A third question concerns its scope: Given that it is characterized as "intuitive," independent of "universal notions," and concerned with "essences," can everything be known by *scientia intuitiva*, or are some

[4]All quotations from the *Ethics* employ the translation of Edwin Curley in Spinoza 1985, except that 'cognition' is used to translate *'cognitio'*. In the main text, citations from the *Ethics* begin with 'E' and the part number, employing the usual system of single-letter abbreviations for 'definition', 'axiom,' 'proposition', 'demonstration', 'scholium', etc. In quotations, Spinoza's own internal citations are left as they occur in the original—i.e., without part numbers where reference is made to an element within the same part of the *Ethics*. 'NS' indicates an addition to the text derived from the *Nagelate Schriften*, the Dutch translation of Spinoza's *Opera Postuma* that appeared in the same year (1677) as the Latin original.

truths beyond its reach? A final question concerns the basis of its value: Given that Spinoza characterizes cognition of both the second and third kinds as "adequate" and "true," why does he nevertheless regard *scientia intuitiva* as the best and most valuable kind of cognition?

## The Nature of *Scientia Intuitiva*

In order to understand how *scientia intuitiva* depends on essences, it is necessary to understand both some familiar features of Spinoza's metaphysics and some less familiar features of his theory of essences.

According to Spinoza, everything is either a substance or a mode of a substance (E1p4d). A substance is both *in* itself[5] and *conceived through* itself (E1d3), whereas the modes of a substance are in and conceived through the substance of which they are modes (E1d5). God (i.e., Nature) is the only substance (E1p14) and has infinitely many *attributes*, each in an infinite or unlimited way (E1d6). These attributes are fundamentally different dimensions (as we might put it) of God's existence and reality, and they constitute God's essence (E1d4). Of these attributes, our minds can conceive only two: Extension and Thought. *Ethics* 1a4 states that "the cognition of an effect depends on, and involves, the knowledge of its cause"; thus, just as whatever is in something must be conceived through it, so too whatever is conceived through something must also be caused by it, and vice versa.[6] Hence, God, which is conceived through itself, must also be self-caused; and the modes of God, which are conceived through God, must also be caused by God. Everything is necessarily caused to be just as it is by God and could not have been otherwise (E1p33). Furthermore, there is a precise parallelism between the causal "order and connection" of things and the causal "order and connection" of the ideas of those things (E2p7). In particular, every mode of Extension is both paralleled by and identical to the idea that is the idea of that mode of Extension (E2p7s); one example of this is the human mind, which is both "the idea of" and "one and the same thing as" the corresponding human body (E2p13). As the idea of the human body, the mind perceives—i.e., has some idea of—"everything that happens in" that body (E2p12).[7] Moreover, every human mind is literally in God as an element in God's infinite intellect (E2p11c), although ideas may have much less power as they are in a human mind in comparison with the power they have as they are in God. Causal relations themselves are attribute-specific: for example, each mode, insofar as it is extended, has only extended causes and effects, whereas insofar as it is thinking, it has only thinking causes and effects (E2p6). Hence, the power of an idea is its "power of thinking"

---

[5]This use of the term 'in' implies inherence, rather than a relation of parts to wholes or spatial inclusion.

[6]Although it is not clear from the formulation of 1a4 whether it is intended as a biconditional or not, Michael Della Rocca (1996, Chapter 1) establishes from its employment that it is so intended. There is no significant distinction between conception and cognition in this context.

[7]For further interpretation of the meaning and significance of this claim, see Garrett 2008.

[*potentia cogitandi*].[8] An *adequate* idea is one that has all of the "internal denominations" (notably, intellectual clarity and distinctness) of a "true idea" (E2d4). A *true* idea, according *Ethics* 1a6, is one that fully agrees with what it represents.

God's modes are of two kinds: *infinite* and *finite*. Infinite modes follow from the "absolute nature" of God's attributes (E1p21-E1p23), doing so either immediately or mediately (in the latter case, by following from other infinite modes that follow more directly from God's attributes). They are therefore pervasive throughout the attributes of which they are modes. Among them are the pervasive features of the attributes that constitute the more general and more specific "laws of nature"—included in what Spinoza in the *Treatise on the Emendation of the Intellect* calls "the series of fixed and eternal things . . . and the laws inscribed in these things as in their true codes, according to which all singular things come to be and are ordered" (TdIE §§ 100–101). Presumably, the more specific laws are caused by the more general laws from which they follow. Finite modes, in contrast, follow not from God's "absolute nature," but rather from God's nature—i.e., essence—insofar as it is modified by other finite modes; thus, finite modes are local and temporary variations in the attributes of which they are modes (E1p28). The infinite individual composed of all finite modes, however, is itself an infinite mode (lemma 7s following E2p13s), one which therefore follows from God's absolute nature. The finite modes are *singular things* (E1p25c)—humans, animals, plants, stars, rocks, and all other things that "are finite and have a determinate existence" (E2d7). These singular things may in turn have their own modes or (as Spinoza often says) "affections."[9]

According to *Ethics* 1d4, God's attributes constitute his essence; but what *is* an essence? In the *Treatise on the Emendation of the Intellect* (TdIE §§ 95–96), Spinoza distinguishes a thing's essence [*essentia*] from its *properties* [*proprium* or *proprietas* indifferently]. The essence of a thing makes it what it is and is captured by a satisfactory definition of it, whereas its properties (in this technical sense, henceforth always intended by the English term) *follow from* this essence without themselves constituting it.[10] Since a thing's properties follow strictly from its essence, the thing can never be without them—in contrast to the thing's merely accidental qualities, which may be either present or absent through the efficacy of external forces.[11] In

[8]For a discussion of power of thinking and its identification by Spinoza with consciousness, see Garrett 2008.

[9]For fuller discussion of the way in which infinite and finite modes follow necessarily from God's nature, see Garrett 1991.

[10]Spinoza's examples are often drawn from mathematics, even though mathematics concerns only "beings of reason." In his example of a circle (TdIE § 96), the essence is being "a figure that is described by any line of which one end is fixed and the other movable"; the properties include "having the lines drawn from the center to the circumference equal." As Spinoza here emphasizes, the understanding of a thing's essence always involves an understanding of how it is or can be caused. In his example of the fourth proportional, having the product of the means be equal to the product of the extremes is a common property of all proportionals, as demonstrated by Euclid.

[11]The distinction among essences, properties, and what I am calling "mere" accidents, is of course a common scholastic one. In scholastic philosophy, the term 'accident' alone can cover both qualities that do and qualities that do not follow from the essence; hence, the latter are often called

many cases, however, things with different essences may have some of the same properties in common. Spinoza applies the distinction between essence and properties throughout the *Ethics*, both to God and to the singular things that are among God's modes. In *Ethics* 1p16 and its demonstration, he writes:

> P16: From the necessity of the divine nature there must follow infinitely many things in infinitely many modes (i.e., everything which can fall under an infinite intellect).
>
> Dem: This Proposition must be plain to anyone, provided he attends to the fact that the intellect infers from the given definition of any thing a number of *properties* that really do follow necessarily from it (i.e., from the very *essence* of the thing); and that it infers more *properties* the more the definition of the thing expresses reality .... (emphasis added)

As this indicates, Spinoza regards all of the modes of God as properties (in the technical sense) that follow from God's essence; not surprisingly, given that everything happens by necessity and nothing is external to God, God has no mere accidents. Since a thing's properties are conceived through its essence, from which they follow, they are also (as Spinoza repeatedly confirms) caused by that essence.[12]

In considering singular things, the *Ethics* discusses two different kinds of essences: the *actual essence* [*essentia actualis*] and the *formal essence* [*essentia formalis*]. The actual essence of a singular thing is the thing's *conatus*, or striving to persevere in its own existence (E3p7). This *conatus*—which serves as the starting point for Spinoza's psychology and political theory, as well as the basis for "teleological" explanation in natural science more broadly—is possessed by every singular thing (E3p6), and it endures for as long as the singular thing endures. A singular thing acts only through its *conatus* (E3p7d), which is expressed in both Extension and Thought, and which may properly be called "appetite" (E3p9s). Through its *conatus*, a singular thing acts to preserve itself both by being the sole cause of its properties and by being a partial cause of other modes both of itself and of other singular things.[13]

The formal essence of a singular thing (sometimes simply "the essence," where context makes this clear), on the other hand, is often contrasted with its existence: for example, *Ethics* 1p25 characterizes God as the efficient cause of not only the "existence" but also the "essence" of things. As Spinoza indicates at *Ethics* 2p8, there are formal essences even for singular things that do not exist:

> 2P8: The ideas of singular things, or of modes, that do not exist must be comprehended in God's infinite idea in the same way as the formal essences of the singular things, or modes, are contained in God's attributes.

The corollary to this proposition refers to "singular things that do not exist, *except insofar as* they are comprehended in God's attributes" (emphasis added), and its

---

"contingent accidents." However, this latter term has misleading connotations for a necessitarian such as Spinoza, and he does not use it.

[12]For an excellent account of the significance to Spinoza of the idea that causation is fundamentally "immanent" causation (E1p18) in which a thing causes it properties through its essence, see Viljanen 2007.

[13]For further discussion of *conatus* and its role is self-preservation, see Garrett 2002.

scholium compares the way in which such singular things are comprehended in God's attributes to the way in which undrawn or undelineated rectangles are comprehended within a circle. Thus, as commentators have often noted,[14] the formal essence of a thing constitutes in some way the "possibility" or "actualizability" of that thing. Since whatever is contained "in" God's attributes is a mode for Spinoza (E1d5) rather than a substance, it follows that formal essences must themselves be modes; and since the formal essences of singular things do not come into and go out of existence as singular things themselves must do, these formal essences must be infinite modes rather than finite modes. More specifically, the formal essence of a singular thing is the pervasive feature of an attribute that consists in the compatibility of the laws of nature of that attribute with the existence of the thing itself, so that the thing itself can exist at any place where the requisite finite causes may prove to be present.[15]

Spinoza provides a general characterization of what "belongs to" an essence at *Ethics* 2d2:

> To the essence of any thing belongs that which, being given, the thing is necessarily posited and which, being taken away, the thing is necessarily taken away; or that without which the thing can neither be nor be conceived, and which can neither be nor be conceived without the thing.

The distinction between actual and formal essences arises, then, because an essence can be "given," and a thing can be "posited," in more than one way. An actual essence is something such that when it is *given as actually existing*, the thing itself is *posited as actually existing*. A formal essence, in contrast, is something such that (i) when it is *given as actually existing*, the thing itself is *posited as possible*; and (ii) when it is *given as instantiated*, the thing itself is then *posited as actually existing*. Since a singular thing actually exists if and only if its actual essence does, we may also think of the actual essence of a singular thing as itself being the *actualization* or instantiation of the thing's formal essence, and hence as that which renders the thing itself actual. Thus, the instantiation of the formal essence of a singular thing produces the singular thing *by* constituting that singular thing's actual essence.[16]

In the case of a singular thing, as we have seen, there is an important distinction between its formal *essence* and its actual *existence*; this distinction collapses, however, in the case of God, for "God's existence and his essence are one and the same"

---

[14]See, for example, Donagan 1973 and 1988, Delahunty 1985, and Matson 1990.

[15]See Garrett 2009 for a fuller argument for this interpretation of formal essences. Insofar as the singular thing is a complex body, its existence and maintenance can be understood as the existence and maintenance of a particular pattern or ratio [*ratio*] of motion and rest ("Physical Digression" following E2p13s).

[16]Spinoza's *Treatise on the Emendation of the Intellect* and his *Short Treatise on God, Man, and His Well-Being* also use the term *essentia objectiva* [objective essence], a term that does not occur in the *Ethics*. An *objective essence* is simply the idea of an essence (specifically, it seems, of a formal essence), so that when an objective essence is *given as existing*, the formal essence of the thing is *posited "objectively"* in the scholastic and Cartesian sense—i.e., posited in Thought.

(E1p20). This is because God's very existence consists in the divine attributes that are also God's essence: Extension cannot be distinguished from the one infinite and necessarily existing extended substance, and Thought cannot be distinguished from the one infinite and necessarily existing thinking substance. Likewise, the distinction between the two *kinds* of essences, actual and formal, collapses in the case of God. Since God's power for existing is infinite and cannot encounter any external obstacle, it is perhaps improper to characterize God as having a *conatus* or "striving" for existence at all; but in any case, the infinite power by which God exists is the same as the infinite power by which God acts, and this power, too, consists precisely in the attributes themselves (E1p34 and E1p34d). Because Extension *is* the extended substance and Thought *is* the thinking substance for Spinoza, God's attributes are also their own instantiations.

We are now in a position to see how *scientia intuitiva* depends on essences. It begins, as Spinoza says, from "an adequate idea of the formal essence of certain attributes of God"—that is, from an adequate idea of God's essence itself as expressed in one or both of the attributes that we can comprehend as constituting that essence. (It is not entirely clear why Spinoza uses the seemingly redundant phrase 'formal essence of certain attributes of God' in *Ethics* 2p40s2. Presumably the formal essence *of* God's essence, if there is such a thing, is just that very essence itself. Perhaps he was striving awkwardly to include both the term 'essence' [since reliance on essences is crucial to *scientia intuitiva*] and a reference to the plurality of attributes [since *scientia intuitiva* can be of either Extension or Thought]. In any case, *Ethics* 5p25d says more simply that *scientia intuitiva* begins from "an adequate idea of certain attributes of God.") This idea is always available to the human mind because, according to *Ethics* 2p46, "the cognition of God's eternal and infinite essence which each idea involves is adequate and perfect": any idea of anything requires an adequate idea of the attribute of which that thing is an expression.

From this starting point in adequate cognition of God's self-causing and self-caused essence, *scientia intuitiva* proceeds to an understanding of the effects of that essence, which are God's properties. These properties include those infinite modes—presumably following from the infinite modes that constitute the laws of nature—that are the formal essences of singular things. However, they also include the finite modes, which are all actually existing singular things. To understand an actually existing singular thing, in turn, is to understand that thing's actual essence—i.e., the force that is brought into existence when the thing begins to exist, that constitutes its activity while the thing exists, and that ceases to exist when the thing is destroyed. *Scientia intuitiva* thus proceeds by its nature from adequate cognition of the necessarily-existing divine attributes to adequate cognition of both the formal essences and the actual essences of singular things, essences that follow from and are caused by the divine attributes.

Spinoza's description of the highest kind of perception in the earlier *Treatise on the Emendation of the Intellect* distinguishes two subclasses, one of which is characterized causally: "Finally, there is the Perception we have when a thing is perceived through its essence alone, or through knowledge of its proximate cause"

(TdIE § 19). This is in no way incompatible with the corresponding definition in the *Ethics*. As he later clarifies in the same work, only something in itself and caused by itself—i.e., God—is perceived through its own essence *alone* (TdIE § 92); and this essence is constituted by the attributes that are the starting point of *scientia intuitiva*. When the mind proceeds to grasp the essences of singular things, it does so, as we have seen, precisely by grasping them through their causes in proper causal order—concluding with the proximate (i.e., immediately preceding) cause.

## *Scientia Intuitiva* and Other Kinds of Cognition

*Scientia intuitiva* thus follows the causal order of nature, moving from cognition of causes to cognition of effects, beginning with adequate cognition of an attribute of God.[17] According to *Ethics* 1a4, however, all cognition of things depends on and "involves" cognition of the causes of those things; and according to *Ethics* 2p46, all cognition "involves" an adequate idea of an attribute of God. How, then, is *scientia intuitiva* different from the other kinds of cognition?

The most obvious way in which the highest and lowest kinds of cognition differ from each other is that *scientia intuitiva* is always adequate and true, whereas opinion or imagination is always inadequate and false. An idea is inadequate in a given mind if and only if the causes of that idea are fully within that mind, so that the mind is its adequate cause. As Spinoza explains it in the *Ethics* (E2p13-E2p30), all imagination—which includes sensory perception—consists fundamentally in awareness of a state of one's own body. (In the specific kind of opinion or imagination that is cognition through signs, one bodily state and its correlative idea reproduce, respectively, another bodily state and its correlative idea with which they have become associated.) However, in being aware of a state of one's body, one is also indirectly and confusedly aware of the various external causes of that state of one's body; when many such objects have contributed to the bodily state, one will be indirectly aware of them all together in a confused way. All ideas as they are in (and thought by) God are adequate and true (E2p32), with the ideas of causes producing ideas of their effects—just as they do in human minds experiencing *scientia intuitiva*. Since the external causes of a human being's bodily states are not themselves in the human body, however, ideas of these bodily states are not adequate and true in the human mind either, even though they are so, of course, in God.

How is this compatible with *Ethics* 1a4? The term generally translated as 'involve' is '*involvere*' a technical logical term in late scholasticism, perhaps best translated more specifically as 'implicate'.[18] And while it is true that ideas of effects always depend on and "involve" cognition of their causes—which is why ideas of

[17]This is the key element in the "ordering" interpretation of the third kind of cognition proposed and defended in the very important article Carr 1978. Carr does not, however, offer an interpretation of the essences of singular things that constitute the later elements in this ordering.

[18]See Gabbey 2008 for a discussion of the history and interpretation of this term.

bodily states can also constitute some representation of their causes—it is one thing to say that having an idea of an effect *implicates* (i.e., carries with it, or requires as a precondition) some conception of the cause, and another thing to say that an idea of an effect exists in a given mind because the presence of the idea of the cause *explains* its presence and power in that particular mind. In the case of *scientia intuitiva*, the mind *produces* or *sustains* its ideas of effects from the adequate ideas of their causes that it already possesses, a process that therefore serves to explain the possession of the ideas of the effects by that mind with whatever degree of power they may have. In imagination or opinion, in contrast, the mind has ideas of bodily states simply because the human mind is the idea of, and perceives everything that happens in, the human body; and in having ideas of these bodily states the mind also necessarily has an inadequate subsidiary or derivative conception of the cause, without fully understanding the essence of either the cause or the effect.

Although opinion or imagination can be distinguished from *scientia intuitiva* in terms of the adequacy and truth of its ideas, reason [*ratio*] cannot: like *scientia intuitiva*, reason's ideas are always adequate and true. As we have seen, however, reason does differ from *scientia intuitiva*—and resembles opinion or imagination—for Spinoza in constituting a way in which we "perceive many things and form universal notions." According to *Ethics* 2p40s2, reason arises from "the fact that we have common notions and adequate ideas of the properties of things." The "properties of things," of course, are to be contrasted with their essences. Spinoza explains common notions in *Ethics* 2p37–2p39: they are ideas of features that are "common to all" bodies and are "equally in the part and in the whole."[19] Because the human mind perceives everything that happens in its body, the human mind must include perception of these common features; and because they are equally in the part and in the whole—i.e., fully and not merely partially present wherever they exist—they can only be conceived fully and adequately. (His suggested example of a common notion is that bodies "can move now more quickly and now more slowly.")

Because these common features must be possessed by all bodies, they must themselves be properties of bodies, following from and "involving" their essences.[20] Moreover, because the ideas of these bodies themselves involve the attribute of Extension, the ideas of these common properties must do so as well. Nevertheless, the common notions are in the mind, initially at least, not *because* they are caused to be in the mind by the presence in the same mind of the ideas of essences

[19]Presumably the parallelism of things and ideas guarantees that just as there are common notions concerning all bodies, so too there are common notions that concern all ideas or minds. For simplicity of exposition, however, I will follow Spinoza procedure in this part of the *Ethics* by discussing only the former, leaving the application to the latter as something easily made.

[20]Miller 2004 identifies the common notions with (ideas of) the laws of nature. However, this seems to me to conflate common properties of bodies, which are therefore modes of those bodies, with pervasive features of Extension, which are therefore infinite modes of God. This is not, of course, to deny that there is an intimate relation between these properties and the infinite modes that constitute the laws of nature, whereby the latter explain the former through their causal contributions to the essences of things.

of bodies or ideas of the attribute of Extension; rather, they are in the human mind simply because they must be an element or aspect of any perception of any body whatever, whether one's own or another. Just as the ideas of the attributes provide the starting point for *scientia intuitiva*, the common notions provide a starting point for reason. Precisely because they are common to all things, however, the objects of the common notions cannot themselves constitute the essence of any particular thing (E2p37).

Common notions and adequate ideas of other properties of things allow the mind to "perceive many things" and to form "universal notions" because they allow the mind to represent in a single idea some or all of the many things, with different essences, that share in a certain property and to draw conclusions from these universal representations. Because these representations are adequate, they allow the mind to cognize or infer "without danger of error" (TdIE § 28) *that* something is true; because they do not proceed through the causal order of essences, however, they do not allow the mind to see *how* and *why* it is true. Thus, in Spinoza's case of the fourth proportional, one who follows the demonstration of Euclid knows that it is a common property of proportions that the product of the means equals the product of the extremes, which allows a calculation *that* $x=6$ in $1/2 = 3/x$. But one who has the highest kind of cognition sees through an understanding of the particular ratio expressed by 1/2 that 3/6 is equally an expression of that same unique ratio.

In *Ethics* 5p36s, Spinoza provides another example, one more important to his own philosophical project. There he contrasts the cognition of the human mind's dependence on God that arises from reason with the cognition of it that arises from *scientia intuitiva*. In Part 1 of the *Ethics*, he writes, he had demonstrated (specifically, in E1p25) that all things depend for their existence and their essence on God. Since the human mind is *something*, it follows that it does so as well—dependence on God is therefore a property of the human mind. But because the demonstration relies on this universal property without deriving it from the essence of the human mind in particular, it provides cognition only of the second kind. By Part 5 of the *Ethics* in contrast, he has explained how the essence of the human mind consists in cognition of the human body and how that cognition depends on and is in God. This cognition of the specific character of the mind's essence and of its dependence on God, he claims, constitutes cognition of the third kind:

> Again, because the essence of our Mind consists only in knowledge, of which God is the beginning and foundation (by IP15 and IIP47S), it is clear to us *how* our Mind, with respect both to essence and existence, follows from the divine nature, and continually depends on God.
>
> I thought this worth the trouble of noting here, in order to show by this example how much the cognition of singular things I have called intuitive, or cognition of the third kind (see IIP40S2), can accomplish, and how much more powerful it is than the universal cognition I have called cognition of the second kind. For although I have shown generally in Part I that all things (and consequently the human Mind also) depend on God both for their essence and their existence, nevertheless, that demonstration, though legitimate and put beyond all chance of doubt, still does not affect our Mind as much as when this is inferred from the very essence of any singular thing which we say depends on God.

Just as the *Treatise on the Emendation of the Intellect* adds an explicitly causal element to the description of the highest kind of perception, so too it adds one to the description of the next highest kind of perception, that corresponding to reason:

> Then there is the Perception that we have when the essence of a thing is inferred from another thing, but not adequately. This happens, either when we infer the cause from some effect, or when something is inferred from some universal, which some property always accompanies. (TdIE § 19)

In addition to stating that this kind of perception involves inference from some universal notion or idea of a property, Spinoza here explicitly states that causes can sometimes be inferred from effects[21]—despite the requirement of *Ethics* 1a4 that cognition of effects already "involve" some cognition of causes. This occurs, presumably, when the mind observes or infers the presence of a property and, based on general knowledge of causal relations, infers something about the cause of that property in that case. Although he indicates that this inference may not proceed "adequately," this does not entail that the resulting ideas are themselves inadequate and false ones in his later technical sense; since he claims in the *Ethics* that all ideas constituting reason are adequate and true, he presumably means only that the inference fails to show exactly what the essence of the cause is and *how* that essence produces the effect.

## The Scope of *Scientia Intuitiva*

The term *'intuitiva'* suggests something direct and immediate in some manner; for example, in his *Regulae ad directionem ingenii* Descartes distinguishes between *intuition* and *deduction*, with the former constituted by immediate intellectual apprehension and the latter by a series of connected steps in which each step is an intuition. Moreover, in his description of the example of the fourth proportional in the *Treatise on the Emendation of the Intellect*, Spinoza writes that those who perceive it by the highest kind of perception "see it not by the force of that Proposition [of Euclid] but intuitively, or without going through any procedure" (TdIE § 24).

Spinoza cannot mean by this, however, that no cognitive steps can be distinguished in the highest kind of perception. The definition of the highest kind of perception on which this passage elaborates explicitly allows that it can occur in perceiving something through perception of its proximate cause (TdIE § 19).[22] Furthermore, as G.H.R. Parkinson (1954) has observed, the definition of *scientia intuitiva* in the *Ethics* specifies that it proceeds *from* one cognition (of attributes) *to*

---

[21]This runs directly contrary to the statement of Nadler (2006: 181) that cognition of the second kind proceeds by inference of effects from causes.

[22]As Carr (1978) notes, this passage counts against the suggestion of Curley (1973) that the distinction between reason and *scientia intuitiva* originated as a distinction between an inferential and a non-inferential kind of cognition in the *Treatise on the Emendation of the Intellect* but changed its character in the *Ethics*.

others (of essences of things), and its description in the *Ethics* version of the example of the fourth proportional explicitly characterizes it as atleast partly inferential: "we see this much more clearly because we infer the fourth number from the ratio which, in one glance, we see the first number to have to the second" (2p40s2). Indeed, *Ethics* 2p47s adds concerning God's essence that "we can deduce from this cognition a great many things which we know adequately and so can form that third kind of cognition of which we spoke." In denying that the highest kind of perception involves any "procedure," therefore, Spinoza cannot mean that we are unable to distinguish ordered steps within it—although in a sufficiently powerful mind, these steps might well be taken instantaneously—but rather that these steps, proceeding as they do directly from ideas of essences, will not require any quasi-syllogistic operations with generalizations and their instances.[23] That such a grasp of the things themselves through their essences is intended to supersede inference through universal intermediaries is strongly confirmed by his remarks in the *Short Treatise* that the highest kind of cognition "has no need . . . of the *art* of reasoning" (KV II.1, emphasis added) but is "an immediate manifestation of the object itself to the intellect" (KV II.22).

If *scientia intuitiva* does not operate with universal notions, then it may seem that there must be some general truths—i.e., truths that are not about merely local or changeable matters—that it cannot reach. This impression may be strengthened by Spinoza's description of his example at *Ethics* 5p36s as aiming to show "how much *the cognition of singular things I have called intuitive*, or cognition of the third kind . . . can accomplish" (emphasis added). Largely on the basis of this passage, it has been proposed that *scientia intuitiva*, in the *Ethics* at least, extends only to the essences of singular things and not to the attributes of God (Curley 1973). As Spencer Carr (1978) argues, however, Spinoza's remark does not entail that *scientia intuitiva* is only of singular things, but merely that the example he is giving in *Ethics* 5p36s, concerning the human mind as a singular thing, does not illustrate how much can be accomplished by *scientia intuitiva* that is *not* of singular things. Certainly, all truths are in and known to God; and it would be surprising indeed if there were some general or pervasive truths—and particularly truths about the divine essence itself—that even God could not know by the highest kind of cognition.

Fortunately, however, we are in position to see how God can know all general and pervasive truths by *scientia intuitiva*. For God's knowledge simply follows the causal order of nature, beginning with an adequate idea of the essence of God—which is self-caused and so understood through itself—and proceeding to adequate ideas of the essences of things. In proceeding to the formal essences of things, *scientia intuitiva* must follow the proper causal order, thereby grasping along the way the infinite modes that constitute the laws of nature. While these laws may properly be characterized as "general" insofar as they are pervasively true and not limited to spe-

---

[23] Parkinson (1954) suggests that *scientia intuitiva* differs from reason in requiring no "application of a rule"; see also Miller (2004, Section 4). However, given that no "universal notions" are involved at all, it seems that a somewhat stronger conclusion, excluding the use of generalizations of any kind, may also be warranted.

cific times and places, Spinoza rejects the view that they are themselves "universals" (i.e., that they are either in nothing or in multiple things); rather they are properties in and of a single substance, God, and as such are present *everywhere*. Thus in the *Treatise on the Emendation of the Intellect* he contrasts infinite modes (i.e., "fixed and eternal things," presumably including both laws of nature and formal essences of things) with "singular changeable things":

> [A]lthough these fixed and eternal things are singular [i.e., particular], nevertheless, *because of their presence everywhere, and most extensive power, they will be to us like universals*, or genera of the definitions of singular, changeable things, and the proximate causes of all things." (TdIE § 101, emphasis added)

On the other hand, the fact that *scientia intuitiva* concerns essences may inspire a somewhat opposite worry about the scope of God's cognition—namely, that (aside from the case of God's own existence, which cannot be distinguished from its essence) it cannot include cognition of the actual existence and changing accidental properties of things. Formal essences, no matter how specific, seem at least in principle to be multiply-instantiable—indeed, instantiated whenever and wherever the requisite finite causes occur.[24] The beginning and ending of existence of a singular thing cannot be determined from its formal essence alone, and each singular thing will have a succession of mere accidents that do not follow from that formal essence either. Fortunately, however, we are now in a position to see how God can know such local and changeable truths by *scientia intuitiva* as well. For the highest kind of cognition extends to the "essences of singular things" *simpliciter*—presumably including not only their formal essences but also their actual essences. In understanding the actual essences of all singular things—singular things that are part of the infinite individual, and which follow from the infinite modes together with other singular things—in their proper causal order and relations, God is also able to grasp how those singular things necessarily come to be, interact with one another, and are then destroyed, permitting *scientia intuitiva* not only of what singular things exist at what times and places, but also what properties and mere accidents they have at each stage in their histories.

God's attributes, God's infinite modes, and God's finite modes (including all the modes and affections of these finite modes) together constitute all that there is to know. In having *scientia intuitiva* of them all, then, God understands everything: not only God's own self-sufficient essence but also how and why each mode follows as it does from that essence. Unfortunately, however, while all truths can thus in principle be known by *scientia intuitiva*, such omniscience is limited to God. For only God has such great power of thinking that it can deduce the actual existence of singular things from the divine essence. At best, human beings are limited to

[24]For this reason, Bennett (1984) argues that ideas of essences, as general natures, cannot constitute thought specifically about particular individuals. But while it is plausible that an idea of a singular thing's *formal essence* alone cannot achieve unique reference to that individual, the parallelism and identity of ideas with their objects guarantees, for Spinoza, the unique reference of each idea to that of which it is properly the idea. In particular, the adequate idea of a singular thing as it is in God achieves in this way unique reference to that singular thing.

*scientia intuitiva* of the divine attributes and the infinite modes (including the formal essences of things). This may be one reason why, in the official definition of *scientia intuitiva* at *Ethics* 2p40s, Spinoza does not explicitly invoke the distinction between actual essences and formal essences: while God can know both kinds of essences by the highest kind of cognition, readers of the *Ethics* can know only formal essences in this way. Human cognition of actual essences—even one's own—is limited to cognition of the first kind.

Human cognition also differs from divine cognition in another way. Human beings generally achieve *scientia intuitiva* as the result of their first achieving cognition through reason. For *scientia intuitiva* begins with adequate cognition of God's attributes; but this adequate cognition is itself in the human mind originally because it is "involved" in the common notions that constitute the starting point of reason; and it is by developing the power of one's thoughts of the common notions that one is able to think of God's attributes themselves with sufficient power of thinking to make significant progress in the highest kind of cognition possible.[25]

Of course, since everything is in God, there are no ideas in human minds that are not also in some way in God. This is true of ideas constituting the lower kinds of cognition as much as it is true of ideas constituting *scientia intuitiva*. But *what* an idea represents is relative to the mind in which it occurs.[26] As the idea of a particular human body is *in God*, it occurs together with adequate ideas of the causes of that human body and its many modes or affections. While this idea of the human body of course "involves" (by E1a4) these other ideas of causes that are also present, the idea considered in itself, need only *represent* the human body itself, which is its proper object [*objectum*]. As this same idea *constitutes a human mind*, however, the ideas of the causes of the bodily affections are external to it, and hence (by E1a4), the ideas in the human mind must also serve *in their own right* to represent their external causes to at least some extent. Similarly, the ideas of properties that constitute the "universal notions" of reason, as they occur in God, represent them *individually* as properties of the particular things from whose essences they follow. As they occur in the less powerful mind of a human being cognizing by reason rather than *scientia intuitiva*, in contrast, they represent all of the things having that property *collectively* and without distinction among the essences involved.

## The Value of *Scientia Intuitiva*

Reason and *scientia intuitiva* both provide cognition that is adequate and true; only opinion or imagination is inadequate and false. Accordingly, both kinds of cognition put matters "beyond all chance of doubt" (E5p36s). Both together lessen the possessor's fear of death (E5p38), and both together constitute the intellect, which is

[25] For an excellent account of the mutually reinforcing character of the two highest kinds of cognition in Spinoza's account of human intellectual development, see Malinowski-Charles 2003.

[26] See Della Rocca 1996 for a compelling argument to this effect.

the "part of the mind that is eternal" (E5p40c). Yet *Ethics* 5p25 ascribes the highest value to *scientia intuitiva* alone: "The greatest striving of the Mind, and its greatest virtue is understanding things by the third kind of cognition." Why is this so?

The demonstration of *Ethics* 5p25 is as follows:

> The third kind of cognition proceeds from an adequate idea of certain attributes of God to an adequate cognition of the essence of things (see its Def. in IIP40S2), and the more we understand things in this way, the more we understand God (by P24). Therefore (by IVP28), the greatest virtue of the Mind, i.e. (by IVD8), the Mind's power, or nature, or (by IIIP7) its greatest striving, is to understand things by the third kind of cognition, q.e.d.

This demonstration infers the greatness of the mind's virtue in having *scientia intuitiva*—which Spinoza equates, via *Ethics* 4d8, with greatness of power—from its constituting a greater understanding of God; and it infers that *scientia intuitiva* constitutes a greater understanding of God, in turn, from *Ethics* 5p24: "The more we understand singular things, the more we understand God." But this itself is initially puzzling: since there is nothing but God and modes of God, it seems that *all* adequate and true cognition—reason as well as *scientia intuitiva*—must be understanding of God.

While reason may also provide understanding of God, however, it will not do so *as powerfully* nor in a way that is as fully *of God* as *scientia intuitiva* does. Since effects, for Spinoza, must acquire whatever power they have from their causes, the total cause of an effect is always more powerful than the effect itself. And since the order and connection of ideas is the same as the order and connection of things, the idea of a total cause is always more powerful than the idea of its effect. But whereas reason proceeds from ideas of properties and from ideas of effects to ideas of causes, *scientia intuitiva* proceeds from ideas of essences, which are the causes of properties, and from ideas of causes to ideas of their effects; hence, the latter is naturally more powerful. Furthermore, in *scientia intuitiva* the mind appropriates within itself the same perfect causal structure of cognition that exists eternally in God: rather than employing universal notions, it allows the mind to understand God and God's modes directly, in just the way that God itself does. Accordingly, *scientia intuitiva* provides not merely truths about God, but direct apprehension of the actual causal structure of God itself.

*Scientia intuitiva* is not merely more powerful cognitively, however; it is also, as *Ethics* 5p36s emphasizes, more powerful emotionally: "it affects our Mind" more. The emotional character of this kind of cognition was important to Spinoza from the beginning: in the *Short Treatise*, the lowest kind of cognition (there called "opinion" [*Waan*]) is identified as the source of passive emotions, the second kind of cognition (there called "belief" [*Geloof*]) is identified as the source of "good desires," and the highest kind of cognition (there called "clear cognition" [*klaare Kennis*]) is identified as the source of "true and genuine love, with all that comes of that" (KV II.2). According to the mature *Ethics*, *scientia intuitiva* produces "the greatest satisfaction of Mind there can be" (E5p27) and gives rise to "the intellectual love of God" (E5p33).

These conclusions of the *Ethics* are straightforward consequences of Spinoza's psychology. *Joy*, according to *Ethics* 3p11s, is a passage to a greater perfection, which is also a greater power for action. *Satisfaction of the mind* or *self-esteem*, by "Definition of the Affects" 25 of the appendix to Part 3 of the *Ethics*, is joy that results from considering oneself and one's own power. *Intellectual love of God*, by *Ethics* 5p32, is joy together with the idea of God as the cause of one's joy. Precisely because *scientia intuitiva* renders one most powerful intellectually, it provides the greatest increase in one's own perfection and power, hence the greatest joy and satisfaction of mind. Because it proceeds from ideas of God's very essence to ideas of God's properties, mirroring and participating in God's own cognition, it not only provides greater joy than mere reason can do; it also identifies *God* as the true *cause* of one's joy far more fully and forcefully than mere reason can do. This intellectual love of God, considered as something eternal in which one participates, is *blessedness* (E5p33s).

As Spinoza understands it, then, *scientia intuitiva* is the best and most powerful kind of cognition because it uniquely mirrors the causal structure of God (i.e., Nature). The fundamental causal structure of God, in turn, is one in which properties follow from essences; while he fully acknowledges and emphasizes the importance of "laws of nature," these have the metaphysical character of properties (specifically, infinite modes) that follow from and are caused by the divine attributes that constitute God's essence. The epistemological merits of *scientia intuitiva* as a kind of cognition are a consequence of its metaphysical correspondence with the divine. As with everything in Spinoza, however, its ultimate value lies in its ethical merits—above all, in its capacity to make its possessor blessed.

## References


Bennett, Jonathan. 1984. *A Study of Spinoza's Ethics* (Cambridge: Cambridge University Press).

Carr, Spencer. 1978. "Spinoza's Distinction Between Rational and Intuitive Knowledge," in *The Philosophical Review* 87: 241–252.

Curley, Edwin. 1973. "Experience in Spinoza's Theory of Knowledge," in Spinoza, edited by Marjorie Grene (Garden City: Anchor).

Della Rocca, Michael. 1996. *Representation and the Mind-Body Problem* (New York: Oxford University Press).

Delahunty, R. J. 1985. *Spinoza* (London: Routledge and Kegan Paul).

Donagan, Alan. 1973. "Spinoza's Proof of Immortality," in *Spinoza: A Collection of Critical Essays*, edited by Marjorie Grene (New York: Anchor).

Donagan, Alan. 1988. *Spinoza* (Chicago: Chicago University Press).

Gabbey, Alan. 1996. "Spinoza's Natural Science and Methodology," in *The Cambridge Companion to Spinoza*, edited by Don Garrett (Cambridge: Cambridge University Press).

Gabbey, Alan. 2008. "Spinoza, Infinite Modes, and the Infinitive Mood," in *Studia Spinozana* 16: 41–66.

Garrett, Don. 1991. "Spinoza's Necessitarianism," in *God and Nature: Spinoza's Metaphysics*, edited byYirmiyahu Yovel (Leiden: Brill).

Garrett, Don. 2002. "Spinoza's Conatus Argument," in *Spinoza: Metaphysical Themes*, edited by John Biro and Olli Koistinen (Oxford: Oxford University Press).

Garrett, Don. 2008. "Representation and Consciousness in Spinoza's Naturalistic Theory of the Imagination," in *Interpreting Spinoza: Critical Essays*, edited by Charles Huenemann (Cambridge: Cambridge University Press).

Garrett, Don. "The Essence of the Human Body and the Part of the Mind that is Eternal," in *The Cambridge Companion to Spinoza's Ethics*, edited by Olli Koistinen (Cambridge: Cambridge University Press).

Malinowski-Charles, Sylvaine. 2003. "The Circle of Adequate Knowledge: Notes on Reason and Intuition in Spinoza," in *Oxford Studies in Early Modern Philosophy* Volume 1, edited by Daniel Garber and Stephen Nadler (Oxford: Clarendon Press).

Matson, Wallace. 1990. "Body Essence and Mind Eternity in Spinoza," in *Spinoza: Issues and Directions*, edited by Edwin Curley and Pierre-François Moreau (Leiden: E. J. Brill).

Miller, Jon. 2004. "Spinoza and the *a Priori*," in *Canadian Journal of Philosophy* 34, 4: 555–590.

Nadler, Stephen. 2006. *Spinoza's Ethics: An Introduction* (Cambridge: Cambridge University Press).

Parkinson, G.H.R. 1954. *Spinoza's Theory of Knowledge* (Oxford: Clarendon Press).

Spinoza, Baruch. 1985. *The Collected Works of Spinoza*, Volume I, edited and translated byEdwin Curley (Princeton: Princeton University Press).

Viljanen, Valtteri. 2007. *Spinoza's Dynamics of Being: The Concept of Power and its Role in Spinoza's Metaphysics*, Reports from the Department of Philosophy No. 19 (Turku: University of Turku).

# *Scientia* in Hobbes

**Douglas Jesseph**

Thomas Hobbes is much better remembered for his political philosophy than his epistemology. Nevertheless, his theory of knowledge is an important contribution to the development of epistemology in the early modern period. The centerpiece of Hobbes's theory of knowledge is his account of *scientia*, or deductively structured systematic knowledge that is grounded in the consideration of causes.[1] The aim of the present essay is to give an overview of Hobbes's account of *scientia*.[2] To that end, I begin with a brief summary of Hobbes's views on knowledge and its kinds, and specifically his distinction between absolute and conditional knowledge. I then examine his treatment of *scientia*, drawing the contrast between it and the less exalted knowledge that is *prudentia* while also examining the dependence on definitions and causes that is characteristic of *scientia*. I close by considering the narrow scope that Hobbes granted to *scientia*, which consists of those areas where we can claim "maker's knowledge" of the things known. Somewhat surprisingly, Hobbes restricts the scope of *scientia* to the sciences of geometry and politics, although it is ultimately unclear whether his theory can do the work Hobbes demanded from it.

D. Jesseph (✉)
University of South Florida, Tampa, FL, USA

[1]Hobbes generally uses the English "science" as the equivalent of the Latin *scientia*. Except when quoting Hobbes's English, I use the Latin term, since the word "science" has acquired a rather different meaning than that intended by Hobbes. In citing Hobbes's works I use the abbreviation "*L*" for *Leviathan* (Hobbes [1651] 2003), followed by chapter and page numbers from the 1651 edition. For Hobbes's *De corpore* (1655) I use his English translation (Hobbes 1656a), which is universally known by its Latin title. References to *De corpore* use the abbreviation "*DCo*" and include part, chapter, and section number separated by full stops. I provide references to the *Opera Latina* (Hobbes [1839–1845] 1966a) and *English Works* (Hobbes [1839–1845] 1966b), using the abbreviations "OL" and "EW".

[2]I am by no means the first to have examined this topic. Useful, if dated, accounts of Hobbes's epistemology and philosophy of science can be found in Peters (1956, Chapters 2–3), Watkins (1965, Chapters 2–3), and Goldsmith (1966, Chapters 1–2). Sorell (1986, Chapters 1–5; and 1996) deals with much of this material, as does Jesseph (1996).

T. Sorell et al. (eds.), *Scientia in Early Modern Philosophy*, Studies in History and Philosophy of Science 24, DOI 10.1007/978-90-481-3077-1_8,

## Hobbes on Knowledge and Its Types

Hobbes's approach to epistemology is dominated by the notion that true knowledge is based on causes, so that to understand something is to be able to explain how it was produced. Causes, in the Hobbesian scheme, can only be mechanical causes, or the motion and collision of material bodies. "Nature", Hobbes declared, "does all things by the conflict of bodies pressing each other mutually with their motions" (Hobbes 1661, Epistle; OL 4: 238). The result is that any claim to knowledge must ultimately involve further claims about the mechanical causes of things. In other words, Hobbes held that having knowledge will typically require that one specify the mechanical process that produces the object of one's knowledge.

Within this general epistemological framework Hobbes distinguished two kinds of knowledge, which he termed knowledge of fact and knowledge of consequence, or (as he also phrased it) absolute and conditional knowledge. The distinction is memorably expressed in *Leviathan*,

> There are of Knowledge two kinds; whereof one is Knowledge of Fact: the other Knowledge of the Consequence of one Affirmation to another. The former is nothing but Sense and Memory and is Absolute Knowledge .... The latter is called Science; and is Conditional; as when we know, that, If the figure shown be a circle, then any straight line through the Center shall divide it into equal parts. And this is the Knowledge required in a Philosopher; that is to say, of him that pretends to Reasoning. (*L* ch. 9, p. 40; EW 3: 71.)

It is significant that when Hobbes called knowledge from sense and memory "absolute" he does not mean that it is unimpeachable or somehow perfect. Rather, the contrast he drew is between knowledge that is "detached in position or relation; independent" and knowledge that is connected to or dependent on other terms or stipulations.[3] In the former case, we can know singular propositions about how things appear to the senses, or how we recall them once to have appeared. In the latter case, we can attain certainty about the link between claims about the world, provided that this link expresses a necessary consequence where one affirmation follows from another.

Absolute knowledge must always be to some extent conjectural, as it depends upon fallible hypotheses about the ways in which motions of external bodies produce sensations in us.[4] We may well enough know that snow appears white, and even that it melts at room temperature. But precisely how snow works to produce a white visual image in our sensory apparatus is a matter of conjecture, to say nothing of how increased temperature causes it to melt. Thus, although knowledge of singular propositions and states of affairs (as well as certain regularities in nature) is

---

[3]The *Oxford English Dictionary* documents uses of "absolute" in this sense from the sixteenth century onward.

[4]Hobbes highlights the necessity of hypotheses in our knowledge of nature in part IV of *De corpore*, where he characterizes "Physiques" as "the finding out by the Appearances or Effects of nature which we know by Sense, some wayes and means by which they may be (I do not say, they are) generated" (*DCo* 4.25.1; *EW* 1: 388).

absolute in Hobbes's sense, such knowledge is not altogether free of some element of hypothesis or conjecture about the causal structure of the world.

## *Scientia*, Signs, and Syllogisms

Hobbes used the terms "science" or *scientia* for conditional knowledge, emphasizing that genuine *scientia* is concerned with the connection between one affirmation and another. The Hobbesian approach to *scientia* is heavily indebted to his conception of language and its role in human reasoning.[5] In fact, he took language to be essential for *scientia*, remarking that "By the Advantage of Names it is that we are capable of Science, which Beasts, for want of them are not; nor Man, without the Use of them" (Hobbes 1684, 25; EW 4; 21).

The origins of *scientia* lie in the imposition of names and making of definitions. Names are either proper or general: a proper name is "singular to one only thing", while a general name is one "Common to many things" (*L* ch. 4, p. 13; *EW* 3: 21). Once names have been imposed, they serve to call to mind the things named, which means that they function either as marks or signs. A mark is a private name used only to remind the speaker of his previous thoughts, while a sign is a name accepted by others in a speech community and used for purposes of communication.

The aim of *scientia* is to employ reason to establish true propositions about the world, a proposition being "a Speech consisting of two Names copulated, by which he that speaketh signifies he conceives the latter Name to be the Name of the same thing whereof the former is the Name; or (which is all one) that the former Name is comprehended by the later" (*DCo* 1.3.2; *EW* 1: 30). Likewise, the truth of a proposition of the form *S* is *P* amounts to the fact that the extension of the subject term *S* falls within the extension of the predicate term *P*.[6]

The reasoning or ratiocination by which true propositions are to be established is a kind of calculation performed on mental contents. As Hobbes boldly proposed, "When a man Reasoneth, hee does nothing else but conceive a summe totall, from Addition of parcels; or conceive a Remainder, from Substraction of one summe from another" (*L* ch. 5. p. 18; *EW* 3: 29). This sort of mental arithmetic is performed most readily by the manipulation of signs or words, but it can be done without them by "adding" or "subtracting" the "phantasms" or sense impressions that are basic to all thought. Hobbes illustrates the case in Part I, Chapter 1, Section 3 of *De corpore* (*EW* 1: 4): suppose a man to see something in the distance, but not distinctly enough to identify it. He thus has only the idea of body, for he knows that this perceived thing must be some visible body or other; on approaching, he discerns that this body

---

[5]On Hobbes's theory of language see Peters (1956, Chapter 5) and Martinich (2005, Chapter 5).

[6]In Hobbes's words: "When two Names are joyned together into a Consequence, or Affirmation; as thus, *a man is a living creature*; or thus, if he be a man, he is a living creature, If the later name *Living Creature*, signifie all that the former name *Man* signifieth, then the affirmation, or consequence is true; otherwise false. For *True* and *False* are attributes of Speech, not of things" (L ch. 4, p. 14; EW 3: 23).

moves itself about "now in one place and now in another", and he can therefore add the idea animated to the earlier idea body. Upon closer examination our investigator "perceives the figure, hears the voice, and sees other things which are signes of a Rationall minde", and therefore frames the idea of rational. The three ideas: body, animated, and rational, are then drawn into a sum when they are applied to the same thing. Thus arises the new idea (namely, the idea of man: a rational, animated body) compounded as a sum out of the three ideas.

Thus described, the mental arithmetic that is reasoning is applied to particular ideas and proceeds without the use of language. However, the more usual case of reasoning will involve language, and the calculation will involve manipulation of words:

> Reason, in this sense, is nothing but Reckoning (that is, Adding, and Subtracting) of the Consequences of generall names agreed upon, for the marking and it signifying of our thoughts; I say marking them, when we reckon by our selves, and signifying, when we demonstrate, or approve our reckonings to other men. (*L* ch. 5, p. 18; *EW* 3: 30)

The drawing of consequences from general names is the concern of logic, and Hobbes modeled his account of *scientia* on classical syllogistic logic.[7] A syllogism, as defined by Hobbes, is "A Speech, consisting of three Propositions, from two of which the third followes" (*DCo* 1.4.1; *EW* 1: 44). Applying his doctrine that reasoning is computation, Hobbes took syllogistic inferences to be a kind of mental addition, where a conclusion is drawn as "a Collection of the summe of two Propositions" (*DCo* 1.4.6; *EW* 1: 48). The aim of all *scientia* is to "establish Universal Rules concerning the Properties of Things", so Hobbes held that the syllogisms at the core of all *scientific* demonstrations need contain only general names, since it is "superfluous to consider any other Mood in direct Figure, besides that, in which all the Propositions are both Universal an Affirmative" (*DCo* 1.4.7; *EW* 1: 49).

The scope of *scientia* is therefore restricted to universal affirmative propositions derived syllogistically from first principles that specify the mechanical causes of things. As we will see, relatively little of what we ordinarily count as knowledge can meet this standard.

## *Scientia* and *Prudentia* Distinguished

The account developed thus far leads to an important distinction between *scientia* and *prudentia*.[8] Both are species of knowledge, but they are distinguished by the fact that *scientia* is certain, infallible, and universal, where *prudentia* is probable, conjectural, and particular. *Prudentia*, which Hobbes equates with accumulated experience, involves the interpretation of natural signs and is something humans share with brute beasts. In contrast, *scientia* requires the imposition of arbitrary linguistic signs and the construction of syllogisms to draw necessary conclusions. So,

---

[7]See Hanson (1990) for more on Hobbes's approach to logic and demonstration.

[8]Barnouw (1990) explores this distinction in greater detail than I can here.

prudential considerations suggest that dark clouds are a likely sign of impending rain, but proper scientific reasoning unerringly demands that a straight line is the shortest distance between two points. As Hobbes puts the matter:

> As, much Experience, is *Prudence*; so, is much Science, *Sapience*.
> For though wee usually have one name of Wisedom for them both; yet the Latines did always distinguish between *Prudentia* and *Sapientia*; ascribing the former to Experience, the latter to Science. But to make their difference appeare more cleerly, let us suppose one man endued with an excellent naturall use, and dexterity in handling his armes; and another to have added to that dexterity, an acquired Science, of where he can offend, or be offended by this adversarie, in every possible posture or guard: The ability of the former, would be to the ability of the latter as Prudence to Sapience; both usefull; but the latter infallible. (*L* ch. 5, p. 22; *EW* 3:37)

The certainty and universal applicability of *scientia* stem from the fact that it is "knowledge of Consequences, and dependence of one fact upon another" (*L* ch. 5, p. 21; *EW* 3:37), whereas *prudentia* is confined to past experience and depends upon a fallible extrapolation from prior cases. In other words, scientific knowledge is conditional knowledge that demonstrates its conclusions from the true causes of the phenomenon to be explained, while prudential knowledge is an instance of absolute knowledge that ultimately rests on fallible assumptions about how the world works.

Another way to see this difference is to note that, although both *scientia* and *prudentia* rely on signs, the signs upon which prudence depends are "but conjectural; and according as they have often or seldom failed, so their Assurance is more or less; but never full and evident "(Hobbes 1684, 20; EW 4: 17–18). In contrast, the signs used in demonstrative reasoning carry "full evidence" with them, and this evidence stems from the fact that they are imposed at will rather than arising from chance correlations found in nature. This is why Hobbes characterized scientia as "conditionall Knowledge, or Knowledge of the consequences of words", which arises when "Discourse is put into Speech, and begins with the Definitions of Words, and proceeds by Connexion of the same into generall Affirmations, and of these again into syllogismes" (*L* ch. 7, p. 30; *EW* 3: 52–53).

## *Scientia* and Causes

The account of Hobbes's epistemology given thus far has a rather significant lacuna, since the connection between *scientia* and causes has been stated but not yet explored. As we have seen, Hobbes insisted that all conditional knowledge must be derived from syllogisms whose premises are either transparently true first principles or derivable from more elementary principles. But a truly scientific (i.e., knowledge-producing) demonstration must employ first principles that express the causes of things. This results in a distinction between two kinds of definitions. One sort of definition (which we can term a stipulative definition) merely indicates how a term is to be used; another (which we can call a generative definition) identifies the cause of a thing.

Both kinds of definitions will be required in the construction of any body of knowledge that can count as scientia. Stipulative definitions are needed to clarify the meanings associated with terms and avoid ambiguity or equivocation. Generative definitions are necessary because genuine knowledge requires causes, and these must be specified among the first principles or definitions.[9]

Although Hobbes did not use the terms "stipulative" and "generative" in distinguishing between these two kinds of definitions, the distinction is nevertheless fundamental to his views on demonstrative knowledge. As he explained:

> Definitions are of two kinds, the first of which merely indicates the nature of the thing, while the second explains the cause or means of generation. But those definitions are the most useful for advancing knowledge which contain the causes and means of generation of the thing defined. This saying of Aristotle's is true: 'To know is to know through causes'; the other definitions, which only declare the essence of the thing defined are usually less fruitful; for nothing follows from them that was not previously contained in them. (Hobbes 1674, 4; *OL*: 4: 156)

When Hobbes speaks of stipulative definitions as those which "indicate the nature", or "only declare the essence" of the thing defined, he means that such definitions simply state the properties we use to sort objects, as when the term man is defined as "rational, animated, body". Such definitions can only tell us how a word is used and cannot be the source of anything more than semantic information. They do not, however, suffice for genuine *scientia* about how the world works. For that, we need to introduce causal definitions.

In some formulations of his epistemology, Hobbes made it seem as if the goal of *scientia* can be realized by analyzing definitions and ordering terms properly, which would require nothing further than stipulative definitions. The universal affirmative propositions appearing in truly scientific syllogisms might seem to depend for their truth only upon speakers' arbitrary conventions about the meanings of general terms, and these can be settled by stipulative definitions. As a result, one might be tempted to think that a true Hobbesian philosopher need tend only to the proper arrangement of terms in his syllogisms. In *Leviathan*, for instance, Hobbes came very close to asserting that the pursuit of *scientia* involves nothing more than the correct ordering of names:

> Reason is not as Sense, and Memory, borne with us; nor gotten by Experience onely as Prudence is; but attayned by Industry; first in apt imposing of Names; and secondly by getting a good and orderly Method in proceeding from the Elements, which are Names, to Assertions made by Connexion of one of them to another; and so to Syllogismes, which are the Connections of one Assertion to another, till we come to a knowledge of all the Consequences of names, appertaining to the subject at hand; and that is it, men call SCIENCE. (*Leviathan* ch. 5. p. 21; *EW* 3: 35)

[9] As Hobbes explained: "The End of Science, is the Demonstration of the Causes and Generations of Things; which if they be not in the Definitions, they cannot be found in the Conclusion of the first Syllogisme that is made from those Definitions; and if they be not in the first Conclusion, they will not be found in any further conclusion deduced from that; and therefore by proceeding in this manner we shall never come to *Science*; which is against the scope and intention of Demonstration" (*DCo* 1.6.13; EW 1: 71).

This appears to make the "Elements" of scientia arbitrarily imposed names, and the reasoning that results in scientia seems to involve little more than investigating the consequences of such names. Taken in the context of Hobbes's insistence on the arbitrariness with which names are "imposed" on the world, this seems to commit him to a strongly conventionalist doctrine in which the truth of a statement will amount to nothing more than speakers' agreement upon definitions and the terms they contain. Hobbes's stronger statements regarding the role of names have misled some interpreters into thinking that he opted for a thoroughly conventionalist philosophy of science. Michael Esfeld, for instance, concludes that in Hobbes's theory of science "there remains systematically only the option of conventionalism: the principles are conventions, about which it is senseless to ask whether they are true" (Esfeld 1995, 131). Ted H. Miller goes further and claims that Hobbes "places limits on truth", by erecting a "barrier" between language and the world, so that knowledge is confined to "the closed circle of propositions and to truths that are strictly formal" (Miller 1999, 157). These "strictly formal" truths are the product of the arbitrary power speakers use to impose names on things by convention. Miller even imagines that the "barrier between Hobbes's philosophy and knowledge of the world" requires that "causal knowledge does not in this case imply knowledge of reality" (Miller 1999, 158). A. P. Martinich claims that Hobbesian definitions "are true by convention; that is, they are true because people choose to use the defined word in the way explained by the definiens" (Martinich 1995, 96). This leads him to see a "tension or contradiction in Hobbes's conception of science" because a strictly conventionalist theory of definition cannot provide definitions that explain how things come to be (Martinich 2005, 165).

Such strongly conventionalist interpretations misconstrue Hobbes's account of *scientia* and its sources.[10] Although stipulative definitions are "purely formal" in the sense that they are arbitrary conventions adopted by a speach community, generative definitions are not. As we have seen, Hobbes held that the aim of truly scientific knowledge is to uncover the (mechanical) causes of phenomena. But the search for such causes necessarily involves more than merely assigning names to things and analyzing definitions. And, of course, Hobbes himself required that a truly scientific definition of something must identify its causes, since "definitions of things, which may be understood to have some cause, must consist of such names as express the cause or manner of their generation" (*DCo* 1.6.13; *EW* 1: 81). Such causal definitions cannot be simple matters of linguistic fiat because it is possible to have a definition that fails to satisfy such a requirement, either by giving no cause of the thing defined or by falsely identifying its causes. To put the point another way: it is

[10] I am not the first to have made this point. Sorell (1986, 41) notes that Hobbes's tendency to stress a "logic book science" dependent only on the analysis of names is counterbalanced by his emphasis on the role of experience and causal principles in his theory of science. Malcolm (2002, 152) makes the important point that "Hobbes's theory of universal truths was a product of his nominalism; and his nominalism was a good deal less extreme than is popularly supposed. He was a nominalist, not an arbitrarist. Hobbes believed that all blue objects, for example, are really similar: our use of the same word to describe them is not a mere freak of human will or fancy."

not a matter of linguistic convention or stipulation whether cyanide causes death in humans, although it is a matter of convention that the English words "cyanide" and "human" refer to the things they do.

## The Scope of *Scientia*

There is an obvious limitation that follows from Hobbes's account of *scientia*: to claim knowledge in the strict sense of *scientia* we must know the causes of things, and it is pretty clear that, at least for a very wide range of things, we cannot guarantee that we have isolated the proper causal story. To return to the example of our (presumed) knowledge that snow is white and melts at room temperature, Hobbes hardly seems to be in a position to fill in the causal account that would explain why and how snow manifests its characteristic properties. And in point of fact, Hobbes was ready to accept that our best theories about nature will have an ineradicably conjectural element because they will be based on fallible hypotheses about the mechanisms of nature. In short, the best we can hope for when it comes to understanding nature is to have absolute knowledge of how things appear to us, and prudential knowledge of what to expect in the future. The result is that we can have absolute knowledge that the stuff we call snow appears white (under standard observation conditions) and has been widely reported to melt at room temperature. Going forward, we can expect that this stuff will continue to appear white and to melt, but this expectation is merely an instance of prudential reckoning that projects past cases into the future without rising to the level of true scientia.

Hobbes accepted this sort of consequence. Indeed, he was prepared to restrict the realm of *scientia* to two disciplines: geometry and politics. These two are the only spheres of human investigation where we can know the causes of things, and as such they are the only two actual instances of scientia. As Hobbes put the matter:

> Of Arts, some are demonstrable, others indemonstrable; and demonstrable are those the construction of the Subject whereof is in the power of the Artist himself; who in his demonstration does no more but deduce the Consequences of his own operation. The reason whereof is this, that the Science of every Subject is derived from a præcognition of the Causes, Generation, and Construction of the same; and consequently where the Causes are known, there is place for Demonstration; but not where the Causes are to seek for. Geometry therefore is demonstrable; for the Lines and Figures from which we reason are drawn and described by our selves; and Civill Philosophy is demonstrable because, we make the Commonwealth our selves. But because of Naturall Bodies we know not the Construction, but seek it from the Effects, there lyes no demonstration of what the Causes be we seek for, but onely of what they may be. (Hobbes 1656b, epistle; *EW* 7: 183–184)

The extent of scientia thus coincides with the range of topics where we can claim "maker's knowledge".[11] Furthermore, *scientia* requires certainty, and Hobbes took certainty to derive from the fact that we have maker's knowledge of the objects of

[11] On maker's knowledge and the role it plays in early modern philosophy (with special emphasis on Francis Bacon) see Perez-Ramos (1988).

genuine *scientia*. Hobbes actually held that "the certainty of all *scientiæ* is equal, or else they would not be *scientiæ*, since to know does not admit of greater or less" (Hobbes 1666, epistle; OL 4: 390). The certainty of *scientia* is thereby purchased at the expense of foregoing any claim to *scientia* about those things we do not ourselves construct, and one might well wonder whether this is a fair epistemological bargain.

There are, in fact, some clear difficulties with Hobbes's approach to *scientia*, not least that it seems to offer no serious response to skeptical doubts about our abilities to acquire knowledge of the external world. If knowledge in a strict, skepticism-immune sense requires an understanding of causes, then (as Hobbes readily admitted) we can know next to nothing about how natural bodies produce their characteristic effects. This is not the place to unravel Hobbes's relationship to skepticism,[12] but his account of *scientia* hardly seems a robust response to any traditional skeptical challenge. Further, the distinction between those things we have caused and those we have not fails to capture the difference between the fallible or conjectural character of natural philosophy and the certainty of geometry and politics.

As Noel Malcolm has observed, the same distinction holds between finding a footprint in the sand (where we are left to conjecture about its origin) and making a footprint oneself (in which case its cause is known). But "the objects of geometry or politics have not just been 'made' by us in this contingent physical sense: they are intentional objects, constituted by the way we think of them" (155). Hobbes's proposal therefore seems to fall short of the goal of providing an adequate account of the nature of *scientia* and its claim to certainty.

In proposing that geometry and politics are the two branches of *scientia*, Hobbes offered the prospect of an account that unifies theoretical and practical sciences. Nevertheless, it seems the scope of *scientia* must also include something beyond simply geometry and politics. What Hobbes termed "first philosophy" or "the making of Definitions, in whatsoever Science they are to be used" (Hobbes 1656b, 4; EW 7: 222) would also be part of *scientia*. It is noteworthy that the definitions that appear in Hobbes's first philosophy include those of body, space, time, and motion. Further, Hobbes makes basic laws of motion, collision, and equilibrium part of first philosophy by placing them in the second part of *De corpore* (which bears the title "The First Grounds of Philosophy") and attempting to demonstrate them.[13] As a result, principles that would ordinarily be classified as part of mechanics or fundamental physics find their way into the foundations of Hobbesian *scientia*.

In the end, Hobbes's theory of *scientia* is an intriguing attempt to maintain the traditional Aristotelian notion that all *scientia* must arise from an understanding of causes while adopting an account of causality that is explicitly mechanistic. The

[12]Tuck (1988a, 1988b) and Popkin (1992a, 1992b) interpret Hobbes as engaged with and reacting to the challenge of skepticism. Sorell (1995) downplays the significance of skepticism in Hobbes.

[13]See Jesseph (2006) on the role of physical and mechanical principles in Hobbes's first philosophy.

resulting hybrid model of *scientia* ultimately failed to win many adherents, and this may well be explained by its rather drastic constraints on what can count as *scientia*. Nevertheless, it remains an interesting example of a seventeenth century epistemological project and one still worthy of our attention.

## References

Barnouw, Jeffrey. 1990. Prudence et science chez Hobbbes. In *Hobbes: Philosophie première théorie de la science et politique*, ed. Yves Charles Zarka and Jean Bernhardt, 103–124. Paris: Presses Universitaires de France.

Esfeld, Michael. 1995. *Mechanismus und Subjectivität in der Philosophie von Thomas Hobbes*. Stuttgart-Bad Canstatt: Fromann-Holzboog.

Goldsmith, M. M. 1966. *Hobbes's Science of Politics*. New York and London: Columbia University Press.

Hanson, Donald W. 1990. The Meaning of 'Demonstration' in Hobbes's Science. *History of Political Thought* 9: 587–626.

Hobbes, Thomas. 1655. *Elementorum Philosophiæ Sectio Prima De Corpore*. London: Andrew Crooke.

Hobbes, Thomas. 1656a. *Elements of Philosophy, the First Section, concerning Body. Written in Latine by Thomas Hobbes of Malmesbury and now translaed in to English*. London: Andrew Crooke.

Hobbes, Thomas. 1656b. *Six Lessons to the Professors of the Mathematiques, One of Geometry, the Other of Astronomy: In the Chaires Set up by the Noble and Learned sir Henry Savile, in the University of Oxford*. London: Andrew Crooke.

Hobbes, Thomas. 1661. *Dialogus Physicus, sive de Natura Aeris Conjectura sumpta ab Experimentis nuper Londoni Habitis in Collegio Greshamensi*. London: Andrew Crooke.

Hobbes, Thomas. 1666. *De Principiis et Ratiocinatione Geometrarum*. London: Andrew Crooke.

Hobbes, Thomas. 1674. *Principia et problemata aliquot geometrica* London: William Crooke.

Hobbes, Thomas. 1684. *Humane Nature: Or The Fundamental Elements of Policy. Being a Discovery of the Faculties and Passions of the Soul of Man, from Their Original Causes; According to such Philosophical Principles as are not commonly known or asserted*. 3rd ed. London: Matthew Gilliflower, Henry Rogers, and Thomas Fox.

Hobbes, Thomas. [1839–1845] 1966a. *Thomæ Hobbes Malmesburiensis Opera Philosophica Quæ Latine Scripsit Omnia in Unum Corpus Nunc Primum Collecta*. Ed. William Molesworth. 5 vols. Reprint, Aalen, Germany: Scientia Verlag.

Hobbes, Thomas. [1839–1845] 1966b. *The English Works of Thomas Hobbes of Malmesbury, Now First Collected and Edited by Sir William Molesworth*. Ed. William Molesworth. 11 vols. Repring: Aalen, Germany: Scientia Verlag.

Hobbes, Thomas. [1651] 2003. *Leviathan, or the Matter, Forme, & Power of a Common-Wealth Ecclesiastical and Civill. A Critical Edition*, ed. G. A. J. Rogers and Karl Schumann. London and New York: Thoemmes Continuum.

Jesseph, Douglas. 1996. Hobbes and the Method of Natural Science. In *The Cambridge Companion to Hobbes*, ed. Tom Sorell, 86–107. Cambridge and New York: Cambridge University Press.

Jesseph, Douglas. 2006. Hobbesian Mechanics. *In Oxford Studies in Early Modern Philosophy*, ed. Daniel Garber and Steven Nadler, vol. 3: 119–152. Oxford and New York: Oxford University Press.

Malcolm, Noel. 2002. Hobbes's Science of Politics and His Theory of Science. In Noel Malcolm, *Aspects of Hobbes*, 146–155. Oxford: Oxford University Press, Clarendon Press.

Martinich, A. P. 1995. A Hobbes *Dictionary*. Blackwell Philosopher Dictionaries. Cambridge, MA and Oxford: Blackwell.

Martinich, A. P. 2005. *Hobbes*. Routledge Philosophers Series. New York and Abingdon, Oxon: Routledge.

Miller, Ted H. 1999. Thomas Hobbes and the Constraints that Enable the Imitation of God. *Inquiry* 42: 149–176.

Perez-Ramos, Antonio. 1988. *Francis Bacon's Idea of Science and the Maker's Knowledge Tradition*. Oxford: Oxford University Press, Clarendon Press.

Peters, Richard. 1956. *Hobbes*. Hammondsworth: Penguin.

Popkin, Richard H. 1992a. Hobbes and Skepticism I. In Richard H. Popkin, *The Third Force in Seventeenth-Century Thought*, 9–26. Leiden and New York: E. J. Brill.

Popkin, Richard H. 1992b. Hobbes and Skepticism II. In Richard H. Popkin, *The Third Force in Seventeenth-Century Thought*, 27–49. Leiden and New York: E. J. Brill.

Sorell, Tom. 1986. Hobbes. The Arguments of the Philosophers, ed. Ted Honderich. London and New York: Routledge and Kegan Paul.

Sorell, Tom. 1995. Hobbes's Objections and Hobbes's System. In *Descartes and His Contemporaries: Meditations, Objections, and Replies*, ed. Roger Ariew and Marjorie Grene, 83–96. Chicago: University of Chicago Press.

Sorell, Tom. 1996. Hobbes's scheme of the sciences. In *The Cambridge Companion to Hobbes*, ed. Tom Sorell, 45–61. Cambridge and New York: Cambridge University Press.

Tuck, Richard. 1988a. Hobbes and Descartes. In *Perspective on Thomas Hobbes*, ed. G. A. J. Rogers and Alan Ryan, 11–42. Oxford: Oxford University Press, Clarendon Press.

Tuck, Richard. 1988b. Optics and Sceptics: The Philosophical foundations of Hobbes's Political Thought. In *Conscience and Casuistry in Early Modern Europe*, ed. Edmund Leites, 235–263. Cambridge: Cambridge University Press.

Watkins, J. W. N. 1965. *Hobbes's System of Ideas: A Study in the Political Significance of Philosophical Theories*. London: Hutchinson.

# John Locke and the Limits of *Scientia*

G.A.J. Rogers

Locke uses the word "science", with the force of the Latin word *scientia*, over twenty times in the *Essay Concerning Human Understanding*.[1] It is a very important concept in his account of knowledge and its limitations and his understanding of it stands in interesting contrast with other major philosophers in the seventeenth century. In many ways his account seems very straightforward but it also contains some perhaps surprising aspects which call into question the picture of Locke as the commonsense philosopher of standard exegesis. Before we reach those latter points, however, we need to outline his central claims and see them in relation to his predecessors and contemporaries.

The two thinkers who stand in clear contrast with Locke and from whom he wished fairly consciously to distance himself are Aristotle and Descartes. Both of these thinkers had ambitious goals for *scientia* and were optimistic about achieving it over a large area of human enquiry of what we now think of as the natural sciences. Locke had considerable admiration for Aristotle—indeed, it is difficult to imagine a philosopher who would not—but he saw major weaknesses, or at least limitations, in Aristotle's account of knowledge.

The crucial document here is Aristotle's *Posterior Analytics*. In that work Aristotle gives us a picture of the sciences which takes it that their correct exposition will be set out in formal axiomatized systems with mathematics, and in particular, geometry providing the paradigm. As we shall see Locke did not quarrel with this ideal. But where he disagreed with Aristotle, or at least his disciples, was in the belief that human beings could realize it. Aristotle, without too much argument, claimed that it was possible to discover the universal truths of nature by induction from particular instances discovered empirically. Locke was quite sure that this was an over-optimistic vision of the human condition. Not only was it unjustified but also it was actually dangerous to think it possible.

G.A.J. Rogers (✉)
University of Keele, Keele, UK

[1]Quotations wil all be taken from the edition of P.H. Nidditch, Clarendon Press, Oxford, 1975, cited by Book, Chapter, Section and page number.

T. Sorell et al. (eds.), *Scientia in Early Modern Philosophy*, Studies in History and Philosophy of Science 24, DOI 10.1007/978-90-481-3077-1_9,

By the time that Locke came to publish the *Essay Concerning Human Understanding* in late 1689 Descartes had been dead for nearly forty years but his influence in the interval had been considerable, perhaps especially in England. Central to Descartes's account of knowledge was the belief that it was possible to establish central truths by pure reason. He claimed that it was possible to discover that the essence of matter was extension and that the essence of mind was thought, purely by reasoning. Locke was quite sure that these and other claims, such as the laws of motion, which Descartes also claimed to establish by pure reason, were unjustified.

Central to Descartes's account of knowledge was a commitment to innate principles in the human mind. The first book of Locke's *Essay* is a total rejection of there being any such source for our knowledge. Like Aristotle Locke held that there were conclusive arguments against the doctrine and he rehearses them in considerable detail. One form of argument in favour of innate knowledge was that there were general maxims which lie at the base of all knowledge which must therefore already be in the mind from inception to account for the knowledge we have. We know immediately, for example, that "tis impossible for the same thing to be and not to be". Against this, Locke held that this is not because it is innate knowledge. Children have no understanding of such sentences when they first hear them "Such kind of general Propositions, are seldom mentioned in the Huts of *Indians*: much less are they to be found in the thoughts of *Children*. . .They are the Language and Business of the Schools, and Academies of learned Nations. . .These Maxims being suited to artificial Argumentation, and useful for Conviction; but not much conducing to the discovery of Truth, or advancement of Knowledge" (I. II.27, p. 64). These "reputed Principles of Sciences" contribute nothing to discovery. "And if *these first Principles* of Knowledge and Science, *are* found *not* to be *innate, no other speculative Maxims can* (I suppose) *with better Right pretend to be so*" (I. II. 28, p. 65). So we see that from outset Locke totally rejects a certain picture of the possibility of *scientia*, that it obtains its status as knowledge from its source in innate principles, truths planted in the mind at conception.

The dominant message that emerges from Locke's account of *scientia* in the *Essay* is how very little of it human beings may expect to obtain. But before we go further into Locke's argument in the body of the *Essay* we would do well to turn to the final chapter, called, appropriately enough, "Of the Division of the Sciences". There are, Locke says, three such divisions, which we may call knowledge of bodies, "The Nature of Things as they are in themselves" and which we now call today the natural sciences, but including the knowledge of spirits, such as angels, God and our own minds. Secondly there is moral philosophy or what we ought to do as rational and voluntary agents. And thirdly, "the Doctrine of Signs", logic and the understanding of language or "Words". These, Locke says, are "the three great Provinces of the intellectual World, wholly separate and distinct one from another" (IV. XXI. 5, p. 721).

Locke's argument in the *Essay* is that our achievements in the area of the first of these, the natural sciences, is very limited. In fact we may well judge that progress in the natural sciences since Locke's day shows that he was over-pessimistic with regard to the potentiality of science to obtain knowledge of the natural world. In fact I shall argue that Locke's pessimism had other causes than purely philosophical ones. Nevertheless, his position anticipates a variety of ones that are to be found in later thinkers in the twentieth century. We also have to bear in mind that Locke's criterion for knowledge was a stronger one than that normally used in everyday claims about the world.

Second, we shall see that Locke is much more sure, on philosophical grounds, that knowledge of ethical truths is within our grasp in a less problematic way than is our knowledge of nature. He also makes the point that moral knowledge is not just theoretical but knowledge of "Right, and a Conduct suitable to it" (IV. XXI. 4, p. 720).

We might begin by considering what Locke is rejecting in his relationship to the possibility of *scientia*. In the chapter "Of the Improvement of our Knowledge" in Book IV Locke contrasts the position that he advocates with the position he rejects:

> It having been the common received Opinion amongst Men of Letters, that *Maxims* were the foundations of all Knowledge; and that the Sciences were each of them built upon certain *præcognita*, from whence the Understanding was to take its rise, and by which it was to conduct it self, in its enquiries into the matters belonging to that Science; the beaten Road of the Schools has been, to lay down in the beginning one or more general Propositions, as Foundations of any Subject. These Doctrines thus laid down for Foundations of any Science, were called *Principles*, as the beginnings from which we must set out, and look no farther backwards in our Enquiries . . . (IV. XII. 1, p. 639).

Locke goes on to say that the reason why this model of knowledge was so widely accepted was because of the paradigm of mathematics, and by mathematics was understood in particular geometry. In this paradigm the schools were of course following Aristotle. In the *Posterior Analytics* he made much use of geometry as a paradigm of knowledge and it was this paradigm that Aristotle wished to see generated for all areas of knowledge. As Jonathan Barnes explains it, the essential thesis of Book A of the *Posterior Analytics* is "that the sciences are properly expounded in formal axiomatized systems. What Euclid later did, haltingly, for geometry, Aristotle wanted done for every branch of knowledge"[2].

Locke was no doubt aware of this Aristotelian aspiration. He had, after all, been taught Aristotle's philosophy as a student and had himself taught it for many years. But there is a sense in which he, like many, perhaps all, of his contemporaries seems to have misunderstood Aristotle's intentions. For, as Barnes argues, Aristotle was not in the *Posterior Analytics* describing a method of discovery but a method of demonstration. Aristotle is not offering a way in which to conduct scientific research it is, rather, "concerned with the organization and presentation of the results of research. . .the primary purpose of demonstration is to expound and render intelligible what is already discovered, not to discover what is still unknown".[3]

We shall see that Locke was indeed guilty of collapsing the distinction between exposition and discovery in his account of contemporary models of philosophical

---

[2]*Aristotle's* Posterior Analytics, Translated with Notes by Jonathan Barnes, Clarendon Press, Oxford, 1975, Introduction p. xi.

[3]Ibid., pp. x–xi.

enquiry but in this respect he was in good company. In the Preface to the French edition of the *Principles of Philosophy* Descartes gives an account of what philosophy is that is identical with the Aristotelian model. Philosophy, he says, includes not only wisdom in our everyday affairs,

> but also a perfect knowledge of all things that mankind is capable of knowing ... In order for this kind of knowledge to be perfect it must be deduced from first causes; thus, in order to set about acquiring it ... we must start with the search for first causes or principles. These principles must satisfy two conditions. First they must be so clear and evident that the human mind cannot doubt their truth ... and secondly, the knowledge of other things must depend on them...[4]

From these principles it will be possible to deduce all knowledge of which mankind is capable. Descartes believes that all men of learning agree with this account of knowledge.

The picture that emerges from Descartes's account of knowledge also fails to draw the distinction between exposition and discovery. And it is clear that he thinks that real knowledge can be obtained purely by deduction from those first principles. It was this holy grail of knowledge that was taken to be the defining goal of *scientia* that Locke was so completely to reject. He takes up the matter immediately after the assault on maxims that we have just considered. He takes it for granted that the paradigm he is rejecting has as its objective the model of obtaining new knowledge that we have just seen was advocated by Descartes. For Locke the Aristotelian schoolmen and the Cartesians, or at least Descartes,[5] were guilty of the same error with regard to the method by which they believed our knowledge could be increased. Locke totally rejected the model: the great advancement of knowledge in the sciences, Locke said, was not owing to principles—the maxims of scholastic logic. He asks rhetorically, "Is it not possible for a young Lad to know, that his whole Body is bigger than his little Finger, but by virtue of this Axiom, that *the whole is bigger than a part*?" (IV. XII. 3, p. 640).

Locke's rejection of "Principles" as a base for knowledge was total and deeply held. At the end of Book I as a part of his assault on innate knowledge he saw their advocates as claiming a kind of moral authority which he totally rejected. He saw the implications in moral and political terms:

> When Men have found some general Propositions that could not be doubted of, as soon as understood, it was, I know, *a short and easy way to conclude them innate*. This being once received, it eased the lazy from the pains of search, and stopp'd the enquiry of the doubtful, concerning all that was once stiled innate: And it was of no small advantage to those who affected to be Masters and Teachers, to make this the Principle of *Principles*, that Principles must not be questioned... (I. IV. 24, p. 101).

---

[4] *Philosophical Writings of Descartes*, translated by John Cottingham, Robert Stoothoff and Dugald Murdoch, Cambridge University Press, Cambridge, 1985, Vol. I, pp. 179–180.

[5] Many of Descartes's followers were aware of difficulties in his model of knowledge even though they accepted much of his account of the natural world.

Once that had been established it provided the conditions for he who decides on the principles to become the "Dictator of Principles, and Teacher of unquestionable Truths; and to make men swallow that for an innate Principle, which may serve to his purpose that teacheth them". The issue of principles, then, was for Locke not simply a matter of logic but went to the heart of his individualist beliefs about the human condition.

What precisely were these individualist beliefs? To appreciate Locke's position we should outline the framework in which his whole philosophy was set. Locke was a committed Protestant Christian like most of his fellow countrymen. He believed that each one of us is set upon earth by God to fulfil his potential and to achieve everlasting happiness in an afterlife. God would not have set us this task if He had not also given to every one of us the capacities to fulfil that end. This end is essentially our carrying out those actions that accord with the will of God and to do that we have to be able to discover what those actions are. Here we can return to Locke's second division of the sciences, his second kind of knowledge: "That which Man himself ought to do, as a rational and voluntary Agent, for the Attainment of any End, especially Happiness" (IV. XXI. 1). This is *Ethicks*. We can, Locke held, come to know what actions are morally right, and we can do this in a variety of ways. As a Christian, Locke held that the most obvious way to acquire such knowledge is by reading the Gospels, but it is not the only route to moral knowledge and it is not one he discusses in the *Essay*, where consideration of specific issues in Christian theology is eschewed.

Another way to reach moral understanding is by rigorous demonstration. To appreciate why Locke saw this as possible requires us to consider the contrast that Locke believed existed between the way in which we can hope to understand the natural world on the one hand and that of the moral order on the other. As an empiricist Locke held that the source of all our ideas is experience. With regard to the natural world our ideas of it must of necessity only be of the surface properties of things. We can never penetrate beneath these properties into the inner structure of physical things. Just as we cannot know much about the bodies in the universe because of their remoteness "there are others that are no less concealed from us by their *Minuteness.* These insensible Corpuscles, being the active parts of Matter, and the great instruments of Nature, on which depend not only all their secondary Qualities, but also most of their natural Operations, our want of precise distinct *Ideas* of their primary Qualities, keeps us in an uncurable Ignorance of what we desire to know about them" (IV. III. 25, pp. 555–556).

Locke did not in principle rule out a science of bodies. He goes on to say that "if we could discover the Figure, Size, Texture, and Motion of the minute Constituent parts of any two Bodies, we should know without Trial several of their Operations one upon another, as we do now the Properties of a Square or a Triangle" (ibid.). It would be as easy to understand why gold dissolves in *aqua Regia* as it is for the locksmith to understand why a key will turn in a given lock. But our condition is, for the most part, not like that. Even when we establish some regularity in our experimental results "whether they will succeed another time, we cannot be certain." (ibid.). So Locke concludes that "how far soever humane Industry may advance useful and *experimental* Philosophy *in physical Things, scientifical* will still be out of our reach" (IV. III. 26, p. 556). Locke goes into considerable detail about the causes of our ignorance of material objects and, for good measure, underlines our even greater ignorance of the spiritual world. In our enquiries of the natural world we can never discover the kinds of necessary connections we can establish in geometry. We can see in our ideas of colour, for example, no natural connection with the impulses which give rise to them and must therefore "ascribe them to the arbitrary Will and good Pleasure of the Wise Architect" (IV. III. 29, p. 560). That there was such a wise architect Locke claimed to know by demonstration. Knowledge of the existence of God was therefore *scientia*, and not a matter of probability.

Locke often returns to underline our limited scope for enquiries into the natural world. The unknown real essences of natural substances makes a search for certainty and knowledge of their natures impossible. I know from experience that, for example, gold has the properties of being yellow, heavy and fusible and take these to define its essence, but then I discover that all gold is malleable and add this property to its essence. But my knowledge reaches only as far as my experience and I have no way of knowing if the list of properties I have so far discovered is all that it has. Indeed, it would be absurd to suppose it was. Of course Locke recognises that a man "accustomed to rational and regular Experiments shall be able to see farther into the Nature of Bodies, and guess righter at their unknown Properties, than one, that is a Stranger to them" But this is "but Judgment and Opinion, not Knowledge and certainty". This, Locke believes, "makes me suspect", "that natural Philosophy is not capable of being made a Science" (IV. XII. 10, p. 645).

This contrasts strongly with his understanding of the position with regard both to knowledge of the existence of God and moral knowledge. He argues that because our natural faculties are not designed to "penetrate into the internal Fabrick and real Essence of Bodies" but are capable of discovering that God exists and enable us to have knowledge of ourselves, "enough to lead us to a clear discovery of our Duty, and great Concernment" (i.e. achieving a place in an eternal afterlife) what human beings should do is to "follow the Direction of Nature". And Locke makes very clear what he takes this to be. It is rational to conclude, he tells us,

> That our proper Imployment lies in those Enquiries, and in that sort of Knowledge, which is most suited to our natural Capacities, and carries in it our greatest interest, *i.e.* the Condition of our eternal Estate. Hence I must conclude, that *Morality* is *the proper Science, and Business of Mankind in general* (IV. XII. 11, p. 646).

Locke makes it quite clear that he is well aware that individuals often have talents which make them distinguished in particular areas, in natural philosophy (Boyle and Newton) and other areas of enquiry (the farmer or sculptor) which they should certainly cultivate for the general benefit of mankind, and he cites the discoveries of iron, printing and the compass as being leading examples of things that have been of enormous benefit to mankind. He does not wish to discourage the study of nature, the benefits of which are, and promise to be, enormous. But he warns against being "too forwardly possessed with the Opinion, or Expectation of Knowledge where it is not to be had; or by ways that will not attain it: That we should not take doubtful Systems, for complete Sciences; nor unintelligible Notions for scientifical Demonstrations" (IV. XII. 12, p. 647).

Why was Locke so sure that knowledge of the kind in question was not to be had? His answer is given by his account of general ideas and general words. In experience all our ideas are particular. We see this apple or that cat. We get to general ideas such as apples or cats, Locke says, by noticing that many of the properties that we see in the particular apple or cat characterise all apples or cats that we have encountered. We then conclude that all apples or cats have these properties and it is in virtue of these facts that they form the natural kinds that they do. This list of properties we take to be the defining characteristics of those objects. But this list is both open ended and uncertain. We do not know that we have ever discovered all the properties of a species. Indeed, we are very sure that there will be a lot of unknown properties to be found amongst all examples of the species. It is this lack of certain knowledge about natural kinds that is at the heart of Locke's pessimism about the possibility of *scientia* in the natural sciences. The properties that we attribute to natural kinds are, for the most part, contingently related one to another, and their discovery can only be made by observation. It is quite different, however, with the nature of mathematical objects. We can by strict logic discover that all triangles have the property of their internal angles equalling two right angles, and the discovered property of the triangle is a necessary property. There could never be a triangle which lacked it.

Locke holds that a similar story can be told about moral truths. Just as all the discourses of the mathematicians "about the squaring of the Circle, conick Sections, or any other part of Mathematicks, *concern not* the *Existence* of any of those Figures: but their Demonstrations, which depend on their *Ideas*, are the same, whether there be any Square or Circle existing in the World, or no" (IV. IV. 8, p. 566). In the same way, moral demonstration has no assumption about existence. Tully's *Offices* are no less true because no society has put his rules into practice or any man lived up to his standards of virtuosity. Locke is quite clear that this does not imply that moral claims are not true or false. If, as he puts it, moral knowledge depends on our own ideas will this not imply that each person can make whatever ideas of morality he pleases? Not so, says Locke, any more than mathematicians can. It can never be right, for example, to call the removal from a man what is his, without his permission, a just act any more than to call an object with four sides a triangle. And as a further, but for Locke ultimate clinching argument, he claims that "where GOD, or any other Law-maker, hath defined any Moral Names, there they have made the Essence of that Species to which that Name belongs; and there it is not safe to apply or use them otherwise" (IV. IV. 10, p. 567).

As is well known, in several places in the *Essay* Locke claimed demonstrative science of morals could be constructed on the model of Euclidian geometry. Nor was Locke the first seventeenth-century philosopher to have that aspiration. At several places in the *Essay* he discusses the possibility and in Book IV he links it directly to his religious beliefs in the way we have already noted. He writes as follows:

> The *Idea* of a supreme Being, infinite in Power, Goodness, and Wisdom, whose Workmanship we are, and on whom we depend; and the *Idea* of our selves, as understanding, rational Beings, being such as are clear in us, would, I suppose, if duly considered, and pursued,

> afford such Foundations of our Duty and Rules of Action, as might place *Morality amongst the Sciences capable of Demonstration:* wherein I doubt not, but from self-evident Propositions, by necessary Consequences, as incontestable as those in Mathematicks, the measure of right and wrong might be made out, to any one that will apply himself with the same Indifferency and Attention to the one, as to the other of these Sciences (IV. III. 18, p. 549).

Although Locke was clearly committed to the possibility of a demonstrative ethics and was encouraged to do so by his Dublin friend William Molyneux he never actually published any such system. There is manuscript evidence that he made some notes which appear to be the beginnings of such a system but they amounted to very little. Other philosophers, including Spinoza, Henry More, and arguably, Hobbes, were more tenacious.

The impossibility of a science of nature was for Locke, ultimately a failure in our capacity to obtain adequate ideas of the natural substances which are the furniture of the universe. It follows that if we were beings with either superior senses—"microscopic eyes"—or an intuitive knowledge of the essence of things, which God, as their creator, would have, then such a science of nature would be available. It was, as we have seen, down to the mediocrity of our natural powers that was the source of our ignorance, not anything intrinsic to the universe at large. And Locke never shared with Leibniz the hope that the secrets of the natural world might reveal themselves to us by means of some super calculation. It is probably true that after Locke that vision was one that never really featured in the history of philosophy again.

# Index



T. Sorell et al. (eds.), *Scientia in Early Modern Philosophy*, Studies in History and Philosophy of Science 24, DOI 10.1007/978-90-481-3077-1,

FSC
www.fsc.org
MIX
Papier | Fördert gute Waldnutzung
FSC® C083411

Kolibri 360 GmbH
Ferdinand-Jühlke-Straße 7
99095 Erfurt, Deutschland
produktsicherheit@kolibri360.de